IMAGES of America

BEDFORD COUNTY

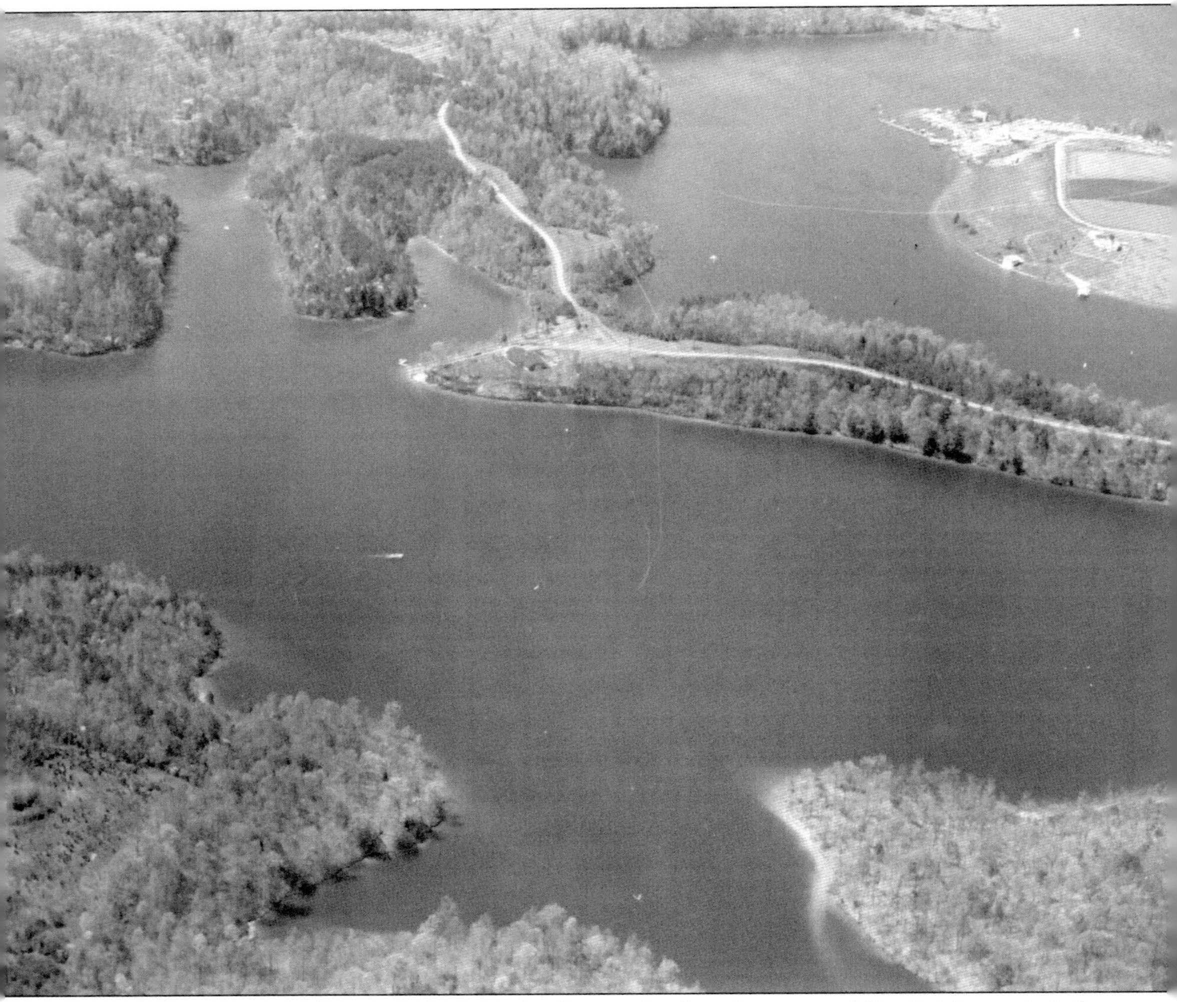

Kirby Grant took this 1973 photograph of an undeveloped portion of Smith Mountain Lake at the Roanoke River channel near Hickory Cove. (Courtesy of Kirby Grant.)

ON THE COVER: This 1920s photograph shows Will Roy Robertson (left), born in 1895, and his brother, Roy, born in 1902. The photograph was taken during an outing near the gap of Smith Mountain, near the area where the Smith Mountain Lake Dam was constructed. The large hill in the background is now two small islands near the dam. (Courtesy of Moorman Robertson.)

IMAGES
of America

BEDFORD COUNTY

Ben Martin

ARCADIA
PUBLISHING

ISBN 978-0-7385-6727-3

Published by Arcadia Publishing
Charleston, South Carolina

Printed in the United States of America

Library of Congress Catalog Card Number: 2008926542

For all general information contact Arcadia Publishing at:
Telephone 843-853-2070
Fax 843-853-0044
E-mail sales@arcadiapublishing.com
For customer service and orders:
Toll-Free 1-888-313-2665

Visit us on the Internet at www.arcadiapublishing.com

This book is dedicated to Harold Martin, Huddleston High School class of 1943 valedictorian, farmer, reader of books, raconteur, history student, and wonderful father.

Contents

ACKNOWLEDGMENTS

I give my most heartfelt and sincere thanks to all of the individuals who contributed photographs for this book; your willingness to provide a piece of your past has allowed me to share our wonderful and collective Bedford County heritage. Thank you, Karen Dillon, for the superb *Roanoke Times* article and for the scores of photograph scans; Candace Nelson and WSLS News Channel 10 for your fine story; Shirley Ring for scouring numerous boxes of photographs; and Gladys Hoover for never being too tired to answer a question. Other notable people and organizations include Dr. James Rucker Jr. for the afternoon you spent searching your family archives; Sam and Debbie Rucker for bringing photographs to me; Frank Ewald for sharing a treasure trove of superb photographs; the Library of Virginia; the Norfolk and Western Historical Photograph Collection; the Virginia Cooperative Extension; Digital Library and Archives, University Libraries, Virginia Polytechnic Institute and State University of Virginia; the Virginia State Dairymen's Association; Linda Owens, assistant director of the Carolina/ Virginia Dairy Products Association; Curtis Baker of Dean Foods; Jay Bryant, Marva Maid Dairy; Allen O'Hara, director of administration and government relations/secretary of Maryland and Virginia Milk Producers Cooperative Association, Inc.; Bobbi Johnson; Richard Bennett, advertising manager, the *Bedford Bulletin*; Kathy Graham, Bedford Area Chamber of Commerce; Georgann Bingham, director of the Traubman Museum of Art; the Honorable Howard Coble, U.S. representative, Sixth Congressional District, North Carolina; John Ewald; Robin Medford; and John Hancock.

INTRODUCTION

"Time flies over us," writes my favorite author, Nathaniel Hawthorne, "but it leaves its shadow behind." My father's favorite book, which he read eight times, was *Gone with the Wind*. During the last summers of his life, he sat often in his front yard on White House Road in Moneta and watched in awe as scores of cars would pass each day down a road that he recalled was once gravel-covered and lonely.

From his perch under the willow tree, he once remarked that he never thought he would live to see a day when there would not be a single mule in the community. Yet, just like Margaret Mitchell's antebellum South, my father's way of life was gone with the wind; the Moneta that he remembered and that I, too, recall as a farming community has been slowly but steadily swept away and engulfed by the larger Smith Mountain Lake community. Even the little hamlet of Moneta was all boarded up by the time he died and resembled a ghost town in a Hollywood Western. Indeed, in 1990, it had been a movie set for Touchstone Pictures' *What About Bob?*

The history of Bedford County has already been well documented. I recommend Volume 3 of *Bedford Villages Lost and Found*, compiled by the Peaks of Otter Chapter of the Daughters of the American Revolution. *A History of Bedford County with Family Histories and Roster of Civil War Soldiers*, reprinted from the 1884 edition of Hardesty's Historical and Geographical Encyclopedia and edited by Peter Viemeister, is another good resource. This list of histories, of course, is not exhaustive.

In the latter work, the first sentence provides the starting point for curious historians: "Bedford county, named in honor, it has been said, of John Russell, the fourth Duke of Bedford, who was the secretary of state of Great Britain from February 13, 1748, to June 26, 1757, was formed in 1753 from Lunenburg county." The former work says, "Archibald Cary obtained a grant for 10,000 acres stretching from present day Bedford to Thaxton and southward around Flat Top Mountain all the way to Goose Creek. His grant was dated 1754 but the survey was made while the area was still a part of Lunenburg County. It bounded other surveys made even earlier."

What a difference 124 years makes! Hardesty's Encyclopedia notes: "Bedford county is thirty-five miles in length, with an average breadth of twenty-five miles, making an area of 776 square miles. The surface is uneven, and in part mountainous. The soil is fertile and produces large crops of wheat, corn, rye, oats, tobacco, and the grasses. Large shipments are made to the market of the cereals, tobacco and cattle. It lies on the eastern base of the Blue Ridge mountains, and is bounded on the north by the James River, and on the south by the Staunton River. The population is 14,557 whites and 10,770 colored—total 25,327. The value of the real estate is $3,744,405.77, the number of acres of land being 489,470, and average of $7.60 per acre. It has 5,000 horses and mules; 11,199 cattle; 6,424 sheep, and 12, 953 swine."

Finally, the manufactories conducted in the county were comprised of 8 tobacco factories, 29 corn and flour mills, and 13 sawmills.

According to Bedford County's official government Web site, www.co.bedford.va.us, the population in 2007 was 66,457. From 1980 to 1990, the population grew 10,729, or 23.50 percent; from 1990 to 2000, it grew 14,715 or 24.37 percent, and since 2000, the number has grown 6,086, or 10.08 percent. The county consists of 764 square miles. It boasts 12 major manufacturing employers, 7 nonmanufacturing employers, and a mean salary of $19,902.46 for packers and packagers to a mean salary of $89,970.94 for general and operations managers. For 2005, the per capita personal income was $32,333.

In *Bedford County*, we will take a more intimate look at the individual people and places that make up Bedford County's past. The construction of Smith Mountain Lake forever altered portions of the landscape of Pittsylvania, Franklin, and Bedford Counties and created many recreation havens along a 500-mile shoreline. Since the 1960s, several generations of immigrants to the lake have contributed vastly to the rich heritage of Bedford County.

Who could have predicted that an old, rickety, *c.* 1900 general store, which sagged mournfully under the weight of its many years in a Southwestern Virginia village, would ever achieve cinematic fame? How did a major motion-picture company ever find this area of Virginia for a film?

Moneta's brief fame, demise, and now rebirth are due to Smith Mountain Lake. This body of water was created by Appalachian Power Company in the early 1960s when a dam was built at Smith Mountain Gap. Though the primary goal of the project was the production of electricity, a recreation destination was obviously expected to be a by-product of the effort. When the $66-million project was completed in 1966, yacht clubs, marinas, public boat launches, an airport, and a state park followed. People came from all over the United States both to relax and to live in Pittsylvania, Franklin, and Bedford Counties, providing corroborating evidence to a claim made in another film, *Field of Dreams*: "If you build it, they will come."

Until the late 1980s, State Route 122 was intersected by the railroad tracks of the old Virginia Railroad. Those tracks essentially cut Moneta in half. Due to increased traffic by a rising population and the danger of fire and rescue squad service being blocked from parts of the community when trains passed, the Virginia State Department of Transportation closed the railroad crossing and constructed a bypass for Route 122 around the little village. The general-store era of Moneta was gone.

This book began as a collection of essays about growing up in the last era of a Bedford County farm community. As I became aware of the Arcadia model of photo essays, however, I realized that my work would not be about my history, but it would focus on our history. I submit this photographic essay, *Bedford County*, so that the county's graveled country roads, horse-drawn plows, and general stores will not be forgotten in the era of the bedroom community. Though it is easy to appreciate the 21st-century downtown Moneta and its new shops, restaurants, and planned communities on its own merits, and though it is not necessary to know that a Coble Milk Transfer Station once stood near the site of the current Lake Christian Ministries, it is a fact that it is nice to know.

As Nathaniel Hawthorne understood well, for some of us, its shadow is still there.

One

The Early Days

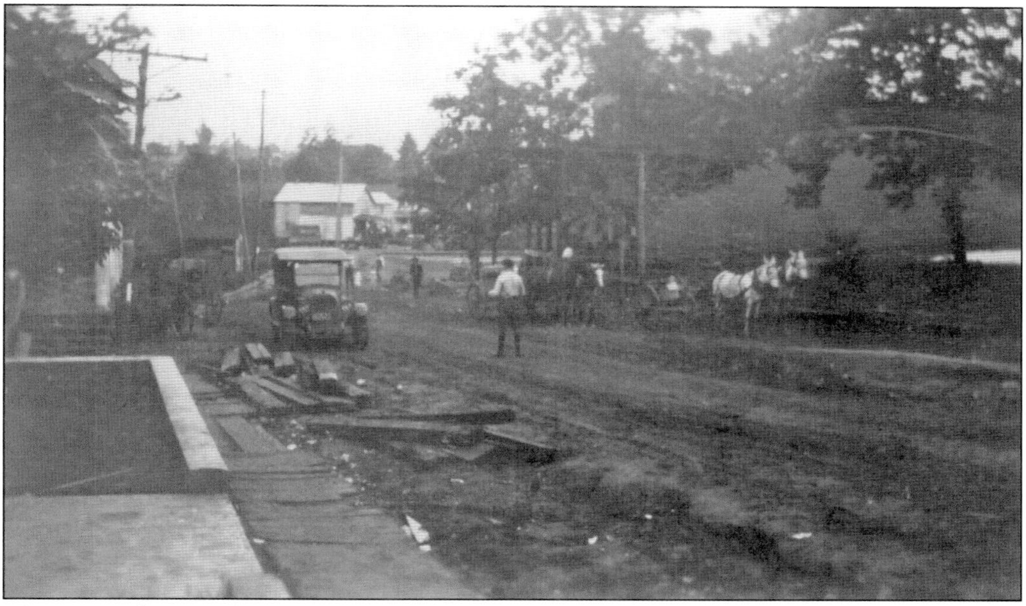

This 1920s photograph shows what would, over a half century later, be known as "Old Moneta," near the railroad tracks of the Virginian Railroad (background). The dirt road would later become State Route 122, around which the Virginia Department of Transportation would build a bypass in the early 1990s. In later years, travelers headed in this direction would be traveling toward Smith Mountain Lake. (Courtesy of Dr. James Rucker Jr.)

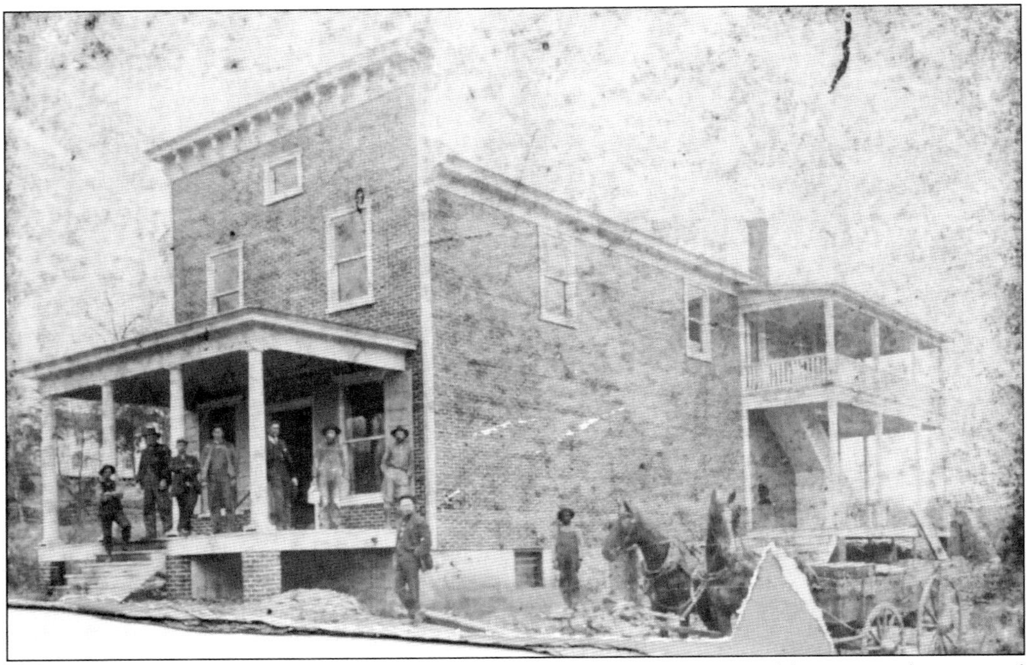

William Howard Meador built this general store in Moneta in 1906. Additions on the original structure's right side followed in subsequent years. The store would be operated by Hanford and Helen Watson and later by Kenneth and Gladys Hoover during its 76 years of operation. This grand old lady of Moneta would have its biggest claim to fame, though, as the general store and coffee shop in the Touchstone Pictures film *What About Bob?* (Courtesy of Gladys Hoover.)

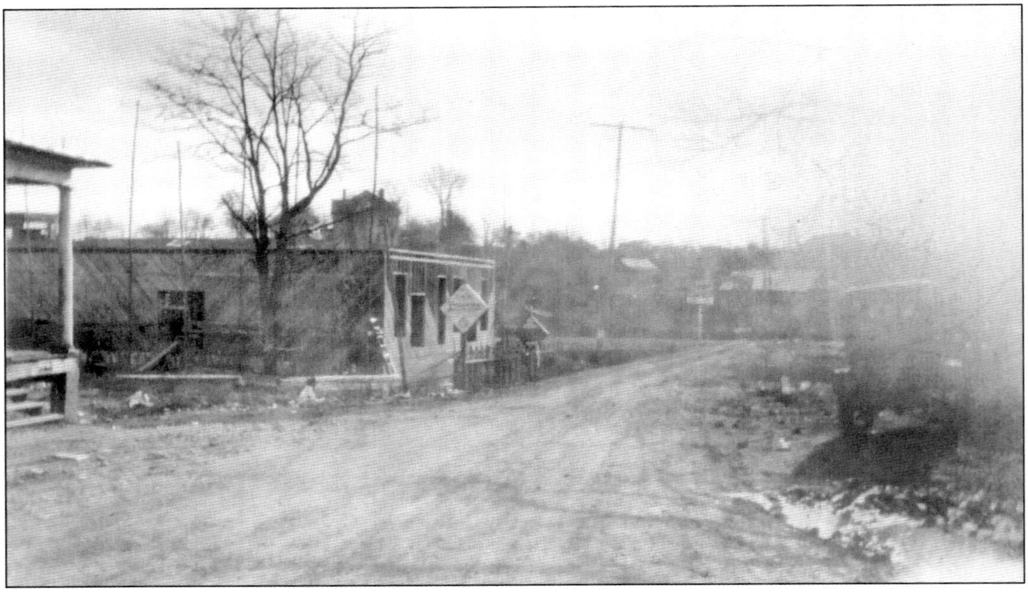

This December 25, 1923, photograph shows the Booker H. Dinwiddie store (left) under construction near the railroad tracks in Moneta. Before it closed in the 1990s, it was known as Moneta Produce. It offered fruits, vegetables, and a vast array of gifts and lake souvenirs to patrons. (Courtesy of Dr. James Rucker Jr.)

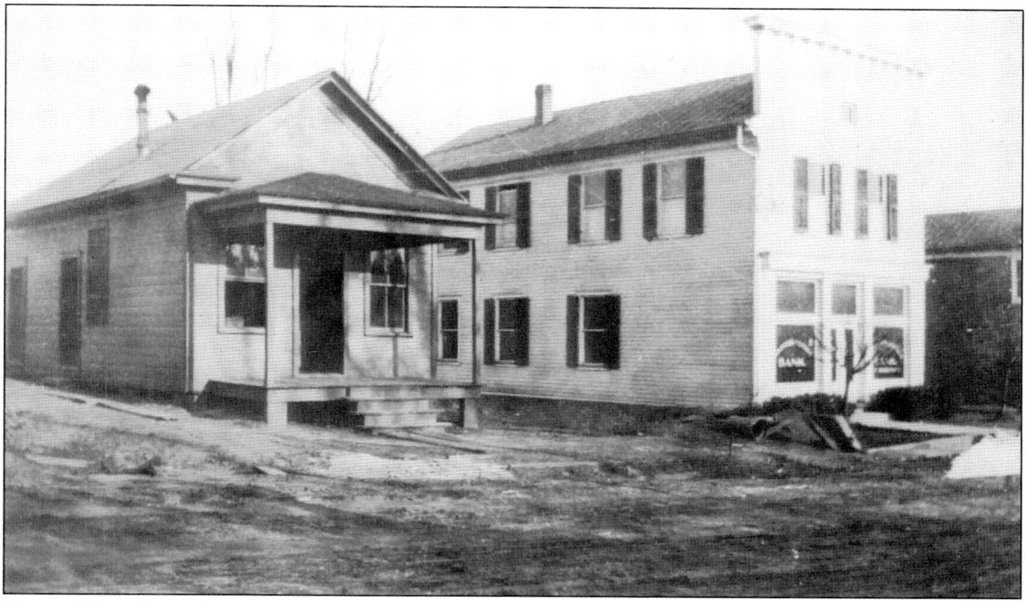

Two of the three buildings in this photograph (center and far left) still stand by the railroad tracks in Moneta. This pre-Depression-era picture shows the Farmers Savings Bank (center), a casualty of October 1929. Both Booker Dinwiddie and his son, Wallace, served as president during its years of operation. Beginning in 1933, it housed the post office. Howard Meador's store is to the far right. (Courtesy of Sam and Debbie Rucker.)

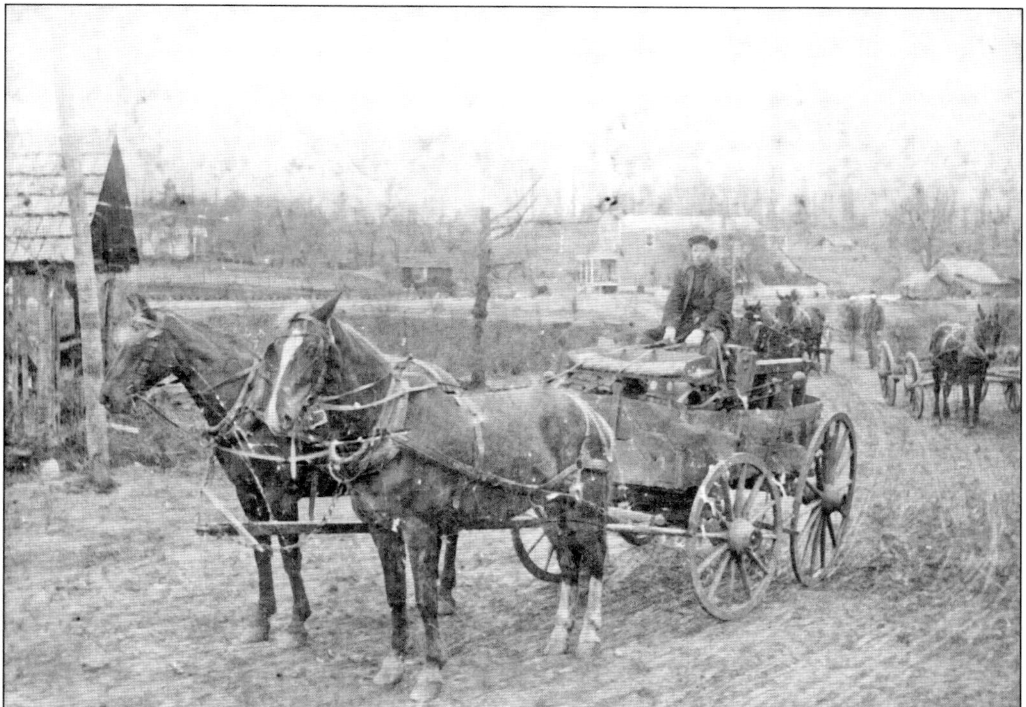

Bob Craghead is crossing the tracks of the Virginian Railroad in Moneta by horse and wagon. The general store that Hollywood made famous can be seen in the background. (Courtesy of Gladys Hoover.)

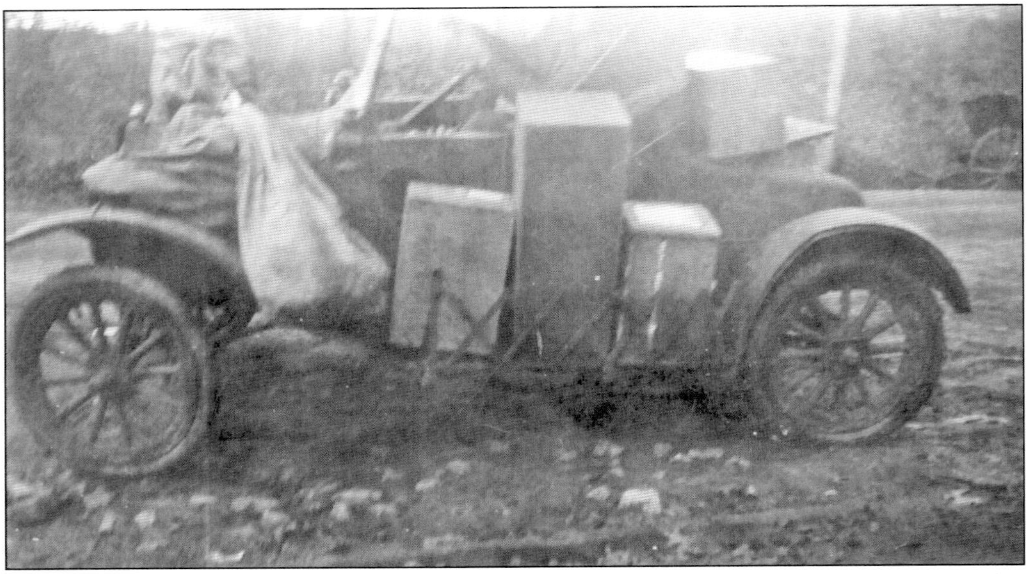

This box-laden automobile belonged to Henry Bond, RFD Carrier No. 1, Moneta, Virginia. This photograph of a much-earlier precursor to today's U.S. Postal Service delivery vehicle was taken December 25, 1923. (Courtesy of Sam and Debbie Rucker.)

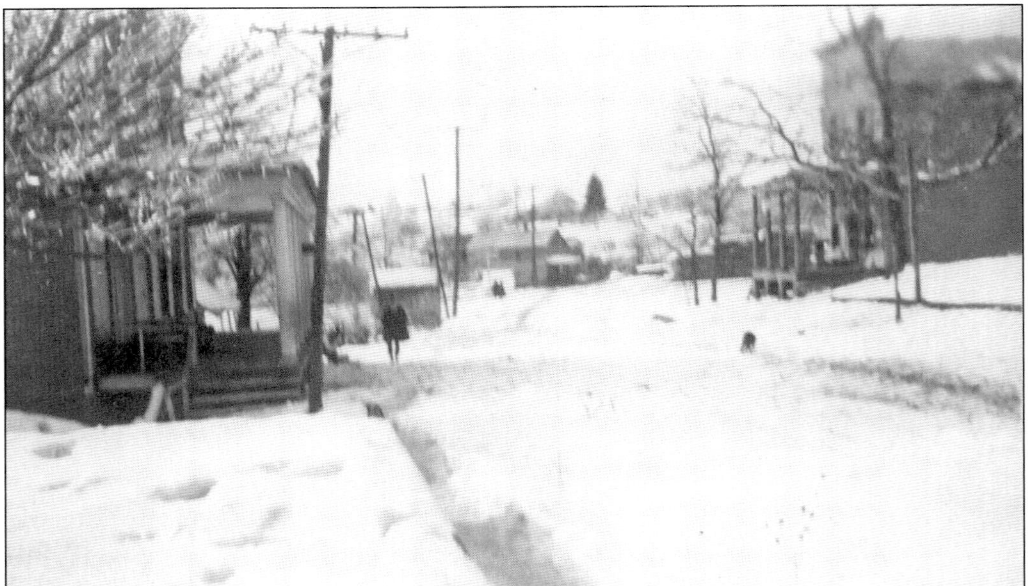

This 1920s photograph of Moneta shows the William Howard Meador store on the left and the David and William Rucker store on the right. Later the Rucker store would be the sight of the first Moneta Fire Department. Before the store's last stint as Sweet's Heating and Air-Conditioning, it served as the Lake Winnipesaukee Fire Department in the Touchstone Pictures film *What About Bob?* (Courtesy of Sam and Debbie Rucker.)

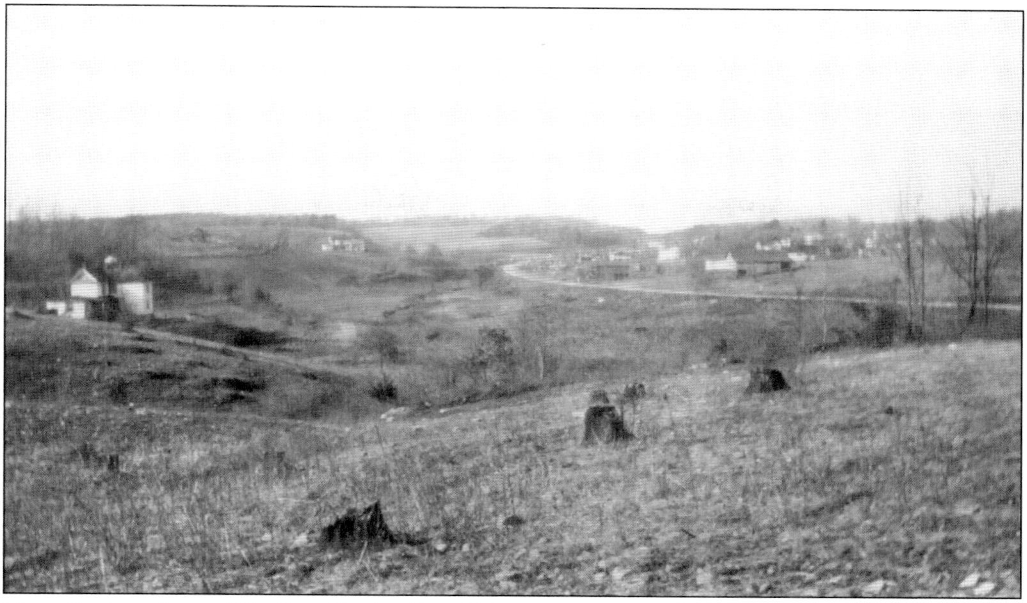

Mary Rucker, sister of Dr. Sam Rucker Jr., of Moneta was trained as a schoolteacher and was also a prolific photographer; most of her photographs are dated and captioned. The caption on this picture reads, "View of Moneta looking from east side." It is dated December 25, 1923. (Courtesy of Sam and Debbie Rucker.)

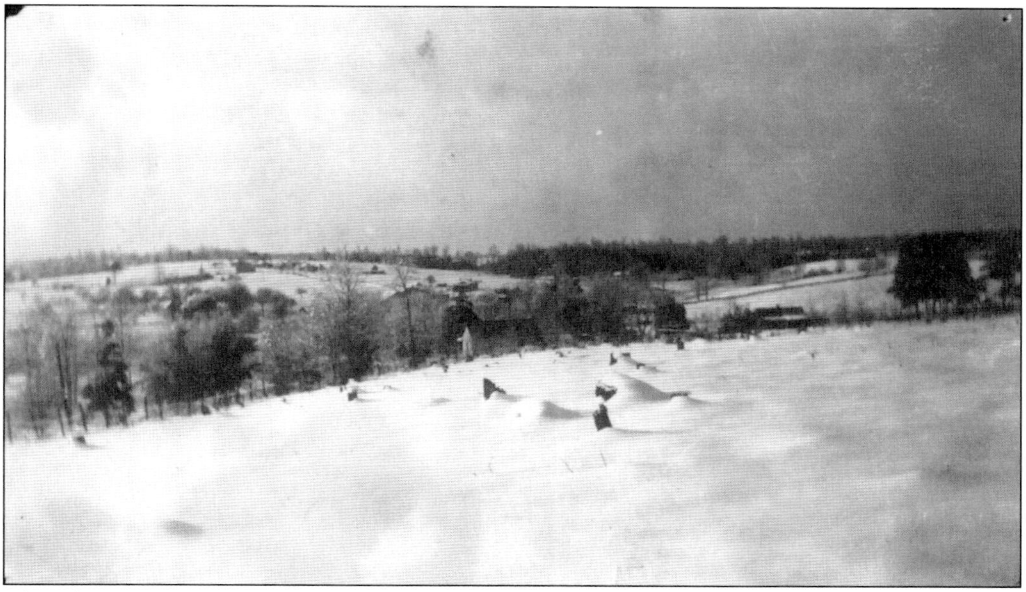

This photograph of Mary Rucker says, "View of Moneta looking south from Rucker House." No date is discernible on this picture. (Courtesy of Sam and Debbie Rucker.)

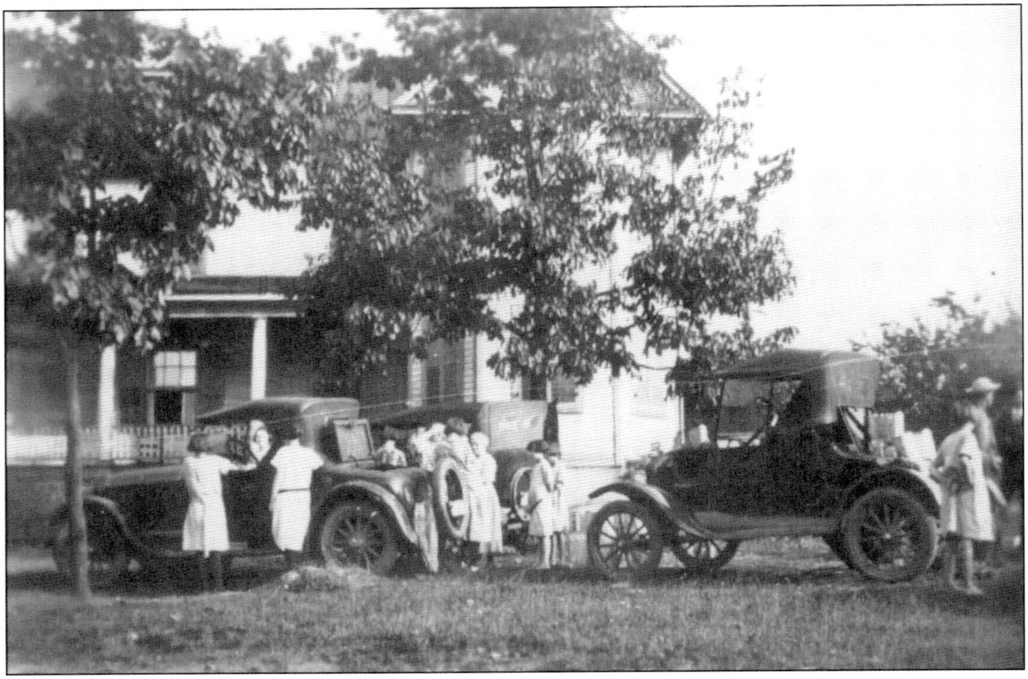

Legend has it that Dr. Sam Rucker Sr.'s hobby was helping federal authorities, known as "revenuers," apprehend moonshine runners. He would be picked up by authorities at the home he built in Moneta in 1909, Rucker House, to be on hand to treat the wounded. One apprehension point was an earlier, covered, wooden Hales Ford Bridge. In this photograph, members of the community are inspecting two confiscated "liquor cars" at Rucker House. (Courtesy of Dr. James Rucker Jr.)

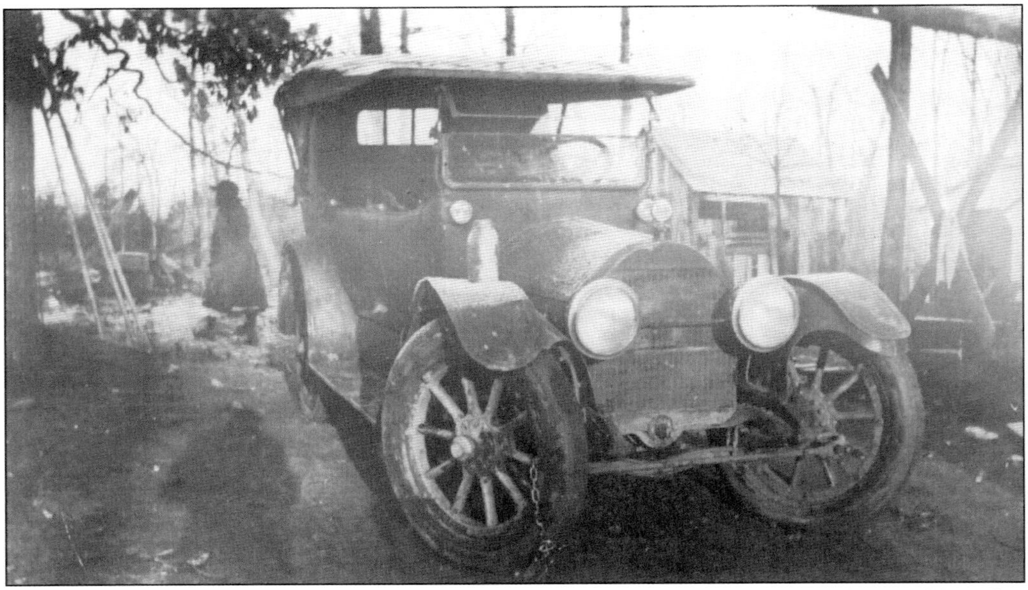

Though Franklin County, Virginia, is known as the "Moonshine Capital of the World," a fair share undoubtedly passed through Bedford County. In this photograph, another confiscated liquor car sits outside Rucker House in the 1920s, proving the adage that "crime doesn't pay." Note the chains on the car's right front wheel. (Courtesy of Dr. James Rucker Jr.)

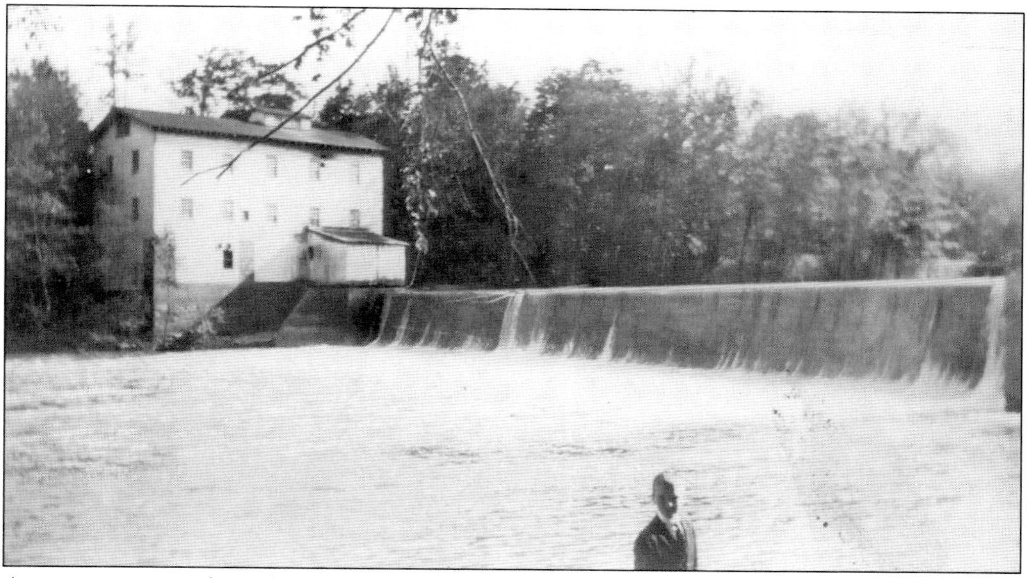

A youngster poses for a photograph at the Davis Mill, a milling operation on Goose Creek that was located near what is now State Route 122 close by the present-day Goose Creek Bridge in Bedford County. A community of Bedford County citizens resided in this area known as the Davis Mill community. (Courtesy of Dr. James Rucker Jr.)

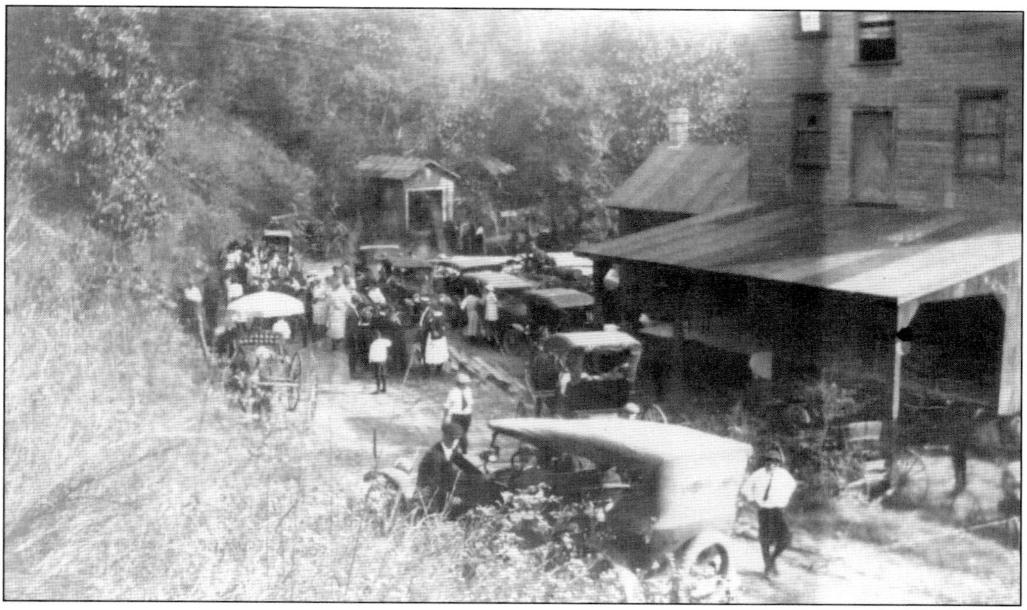

In this 1920s photograph, churchgoers attend a baptism at Davis Mill. (Courtesy of Dr. James Rucker.)

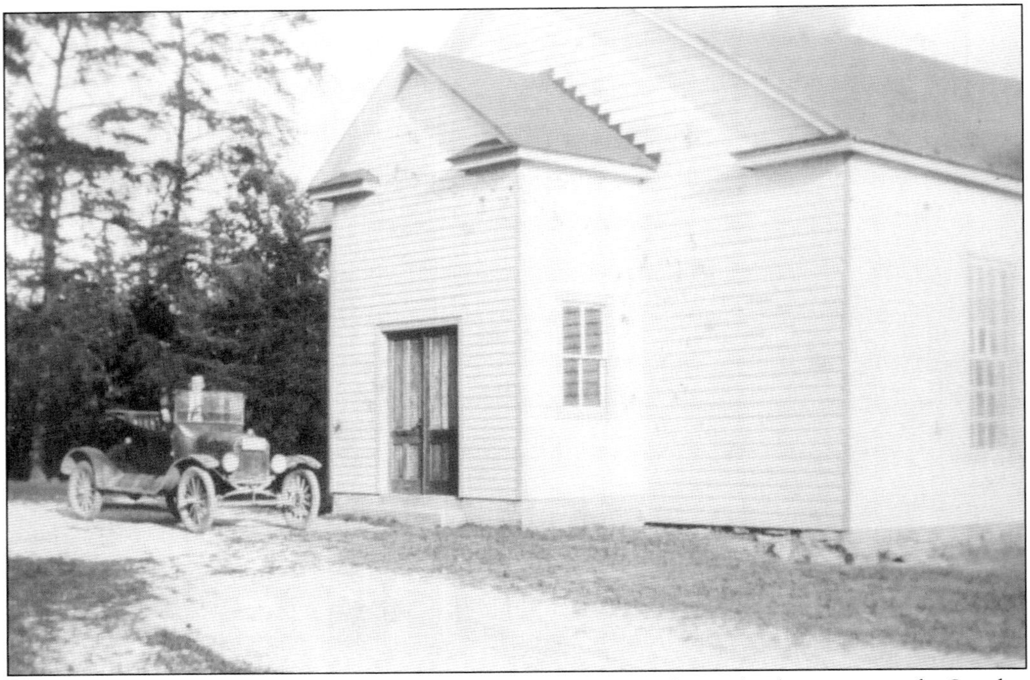

In this undated photograph from the 1920s, this motorist is, perhaps, the first to arrive for Sunday services at Wilson Methodist Church in Body Camp, Virginia, in Bedford County; many of his fellow parishioners would have undoubtedly arrived by an earlier mode of transportation: horse. (Courtesy of Sam and Debbie Rucker.)

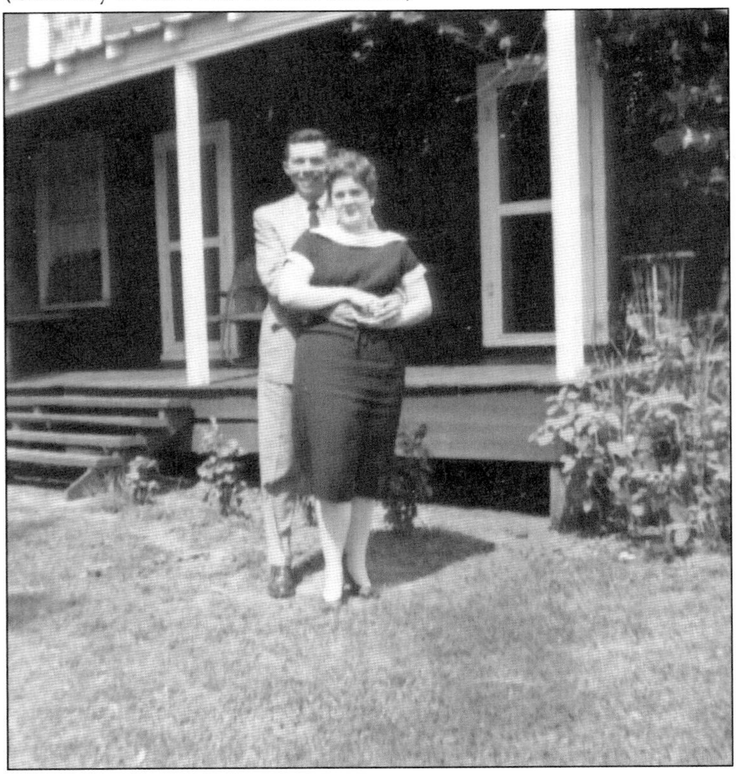

Harold and Ann Martin pose for a picture shortly after their marriage on May 22, 1959, outside what had been Harold's boyhood home. This house was built by John Kasey, who was the magistrate for the Staunton River Magisterial District, in 1870. The Martins renamed the 140-acre farm the Kasey Seats because worship services were held on the property in the late 1800s. The services were held outside, and makeshift pews were fashioned from old boards laid across rocks and tree stumps. (Courtesy of the author.)

A teenage Harold Martin enjoys a nice afternoon outside the Kasey House around 1937. The house and adjoining 99 acres first came into the Martin family in November 1939, when it was sold by Ruby Hackworth to Garland Martin for the sum of $2,500. Martin later sold the farm to his brother, Carlton, who lived here until 1946, the year he sold it to his sons, Harold and Willard. In 1948, Harold Martin became the sole proprietor. (Courtesy of the author.)

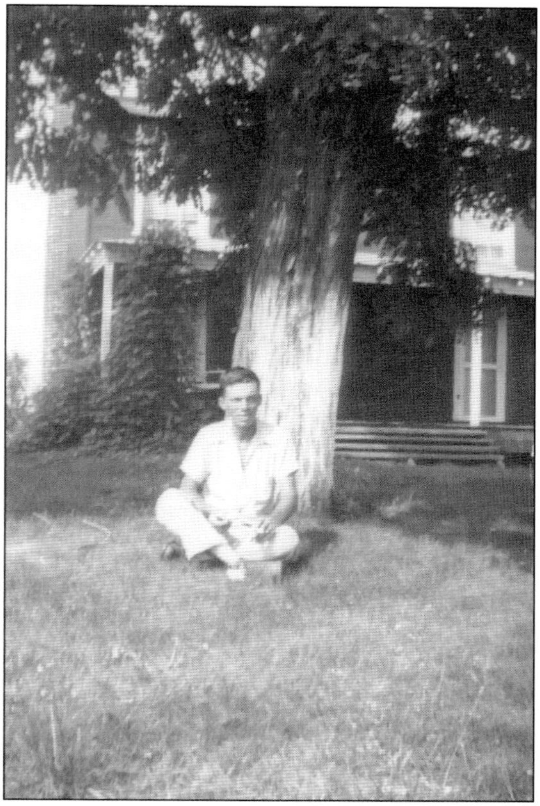

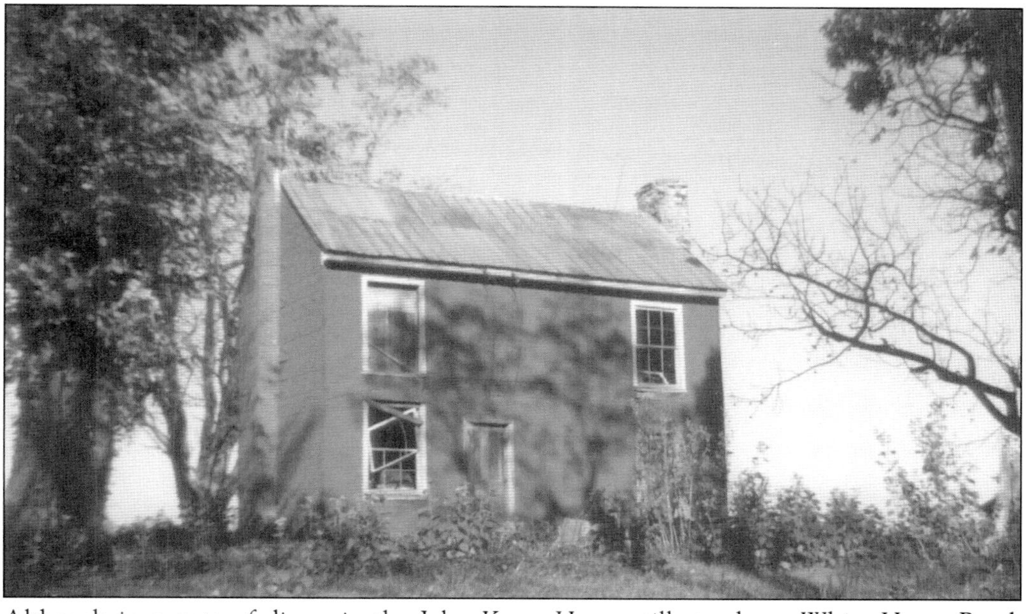

Although in a state of disrepair, the John Kasey House still stands on White House Road. Real estate transactions here show how property values have skyrocketed over the past century: on August 27, 1881, the farm was deeded to Laura Kasey by William Wingfield and Joseph Burroughs for $300. However, on March 14, 1872, Wingfield and Burroughs had purchased the house and 99 acres of land for $200.89. (Courtesy of Robin Medford.)

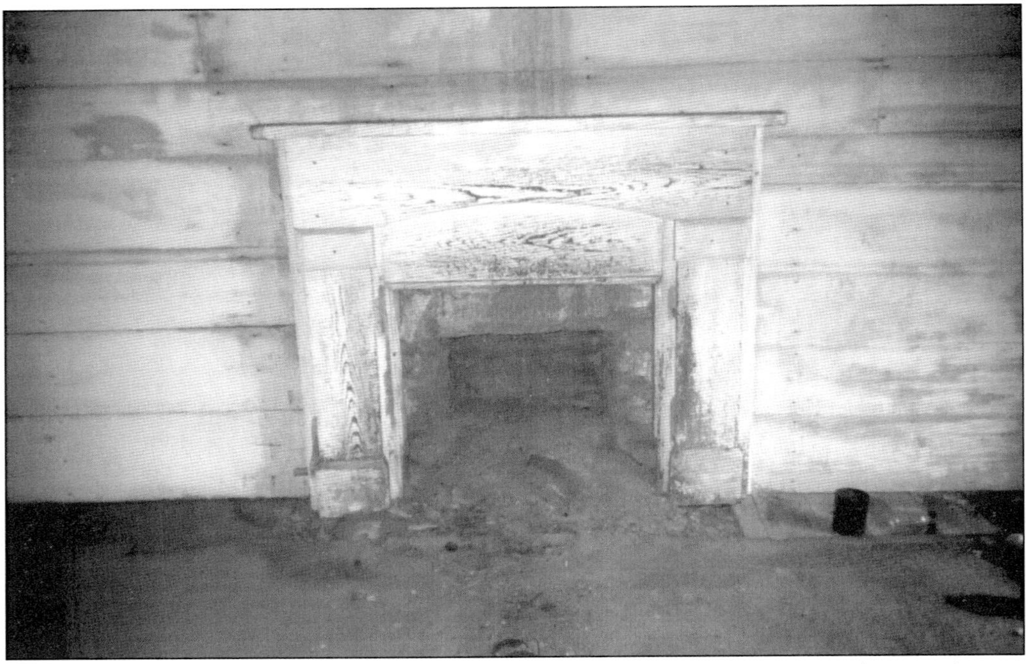

For the tough souls who lived in the Kasey House since 1870, there were neither woodstoves nor elaborate central heating systems. This fireplace, which is located in an upstairs bedroom, provided the only warmth available until woodstoves were added by subsequent owners in the 1900s. Harold Martin once told the author that you needed as many blankets underneath you as you had over you if you were to have any chance of not freezing to death. Jokingly, he once remarked that the homestead was good for keeping the leaves and the rain off of its inhabitants, but it was not good for much else. (Courtesy of the author.)

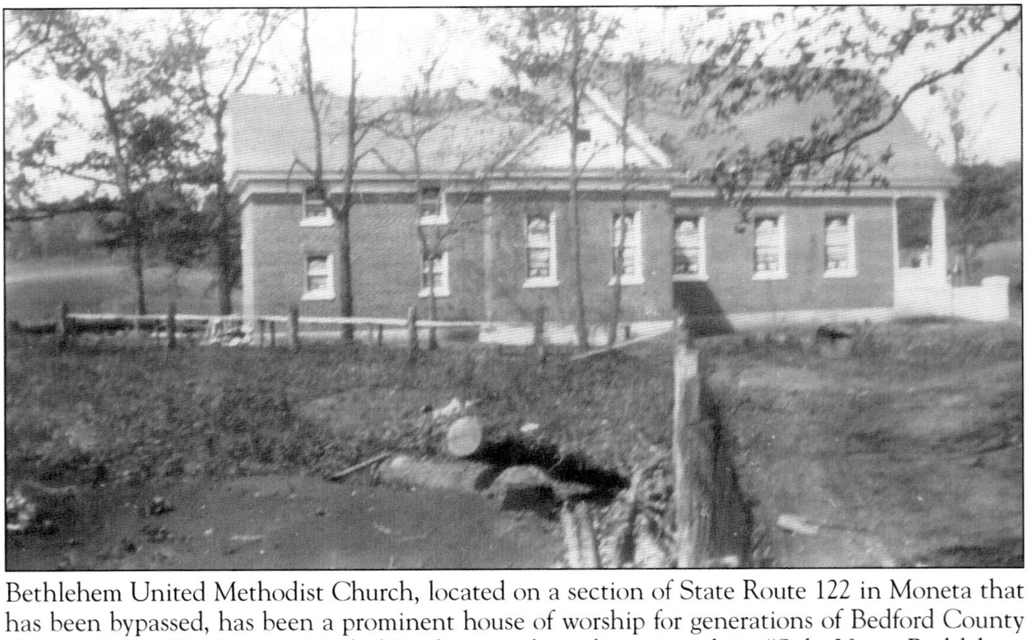

Bethlehem United Methodist Church, located on a section of State Route 122 in Moneta that has been bypassed, has been a prominent house of worship for generations of Bedford County citizens. Mary Rucker captured this photograph and captioned it, "Side View, Bethlehem Church, April 21, 1929." (Courtesy of Dr. James Rucker Jr.)

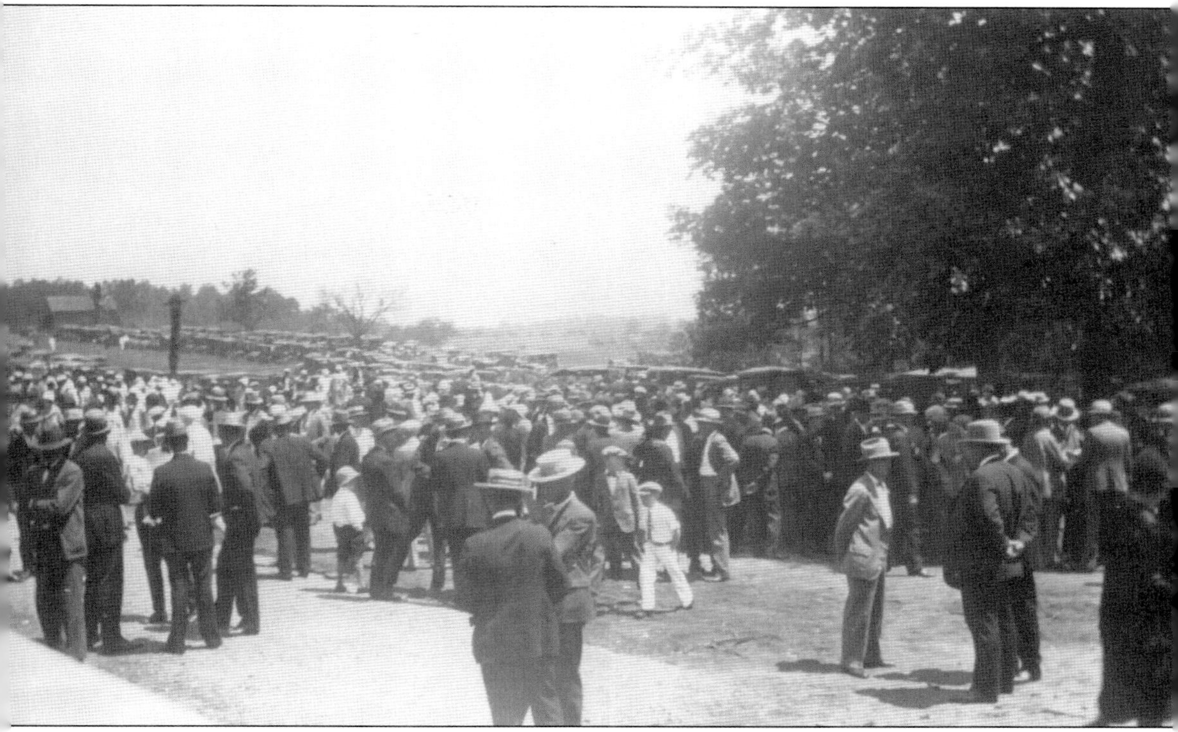

Hundreds of people gather for a dedication service on June 30, 1929, at Bethlehem United Methodist Church. (Courtesy of Dr. James Rucker Jr.)

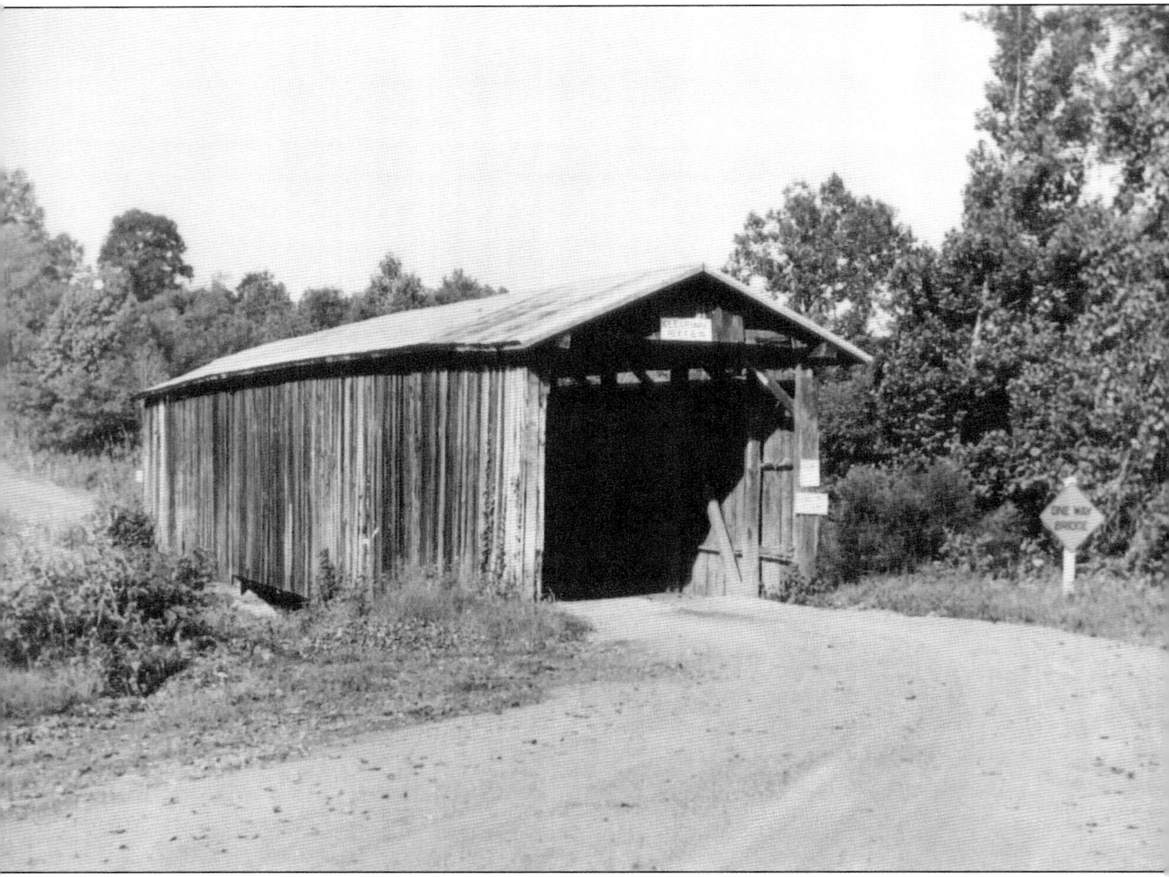

Travelers crossing this covered bridge would have been wise to slow before entering it: the sign "One Way Bridge" provides plenty of incentive. This bridge was on Route 297 and spanned the Little Otter River. (Courtesy of Frank Ewald.)

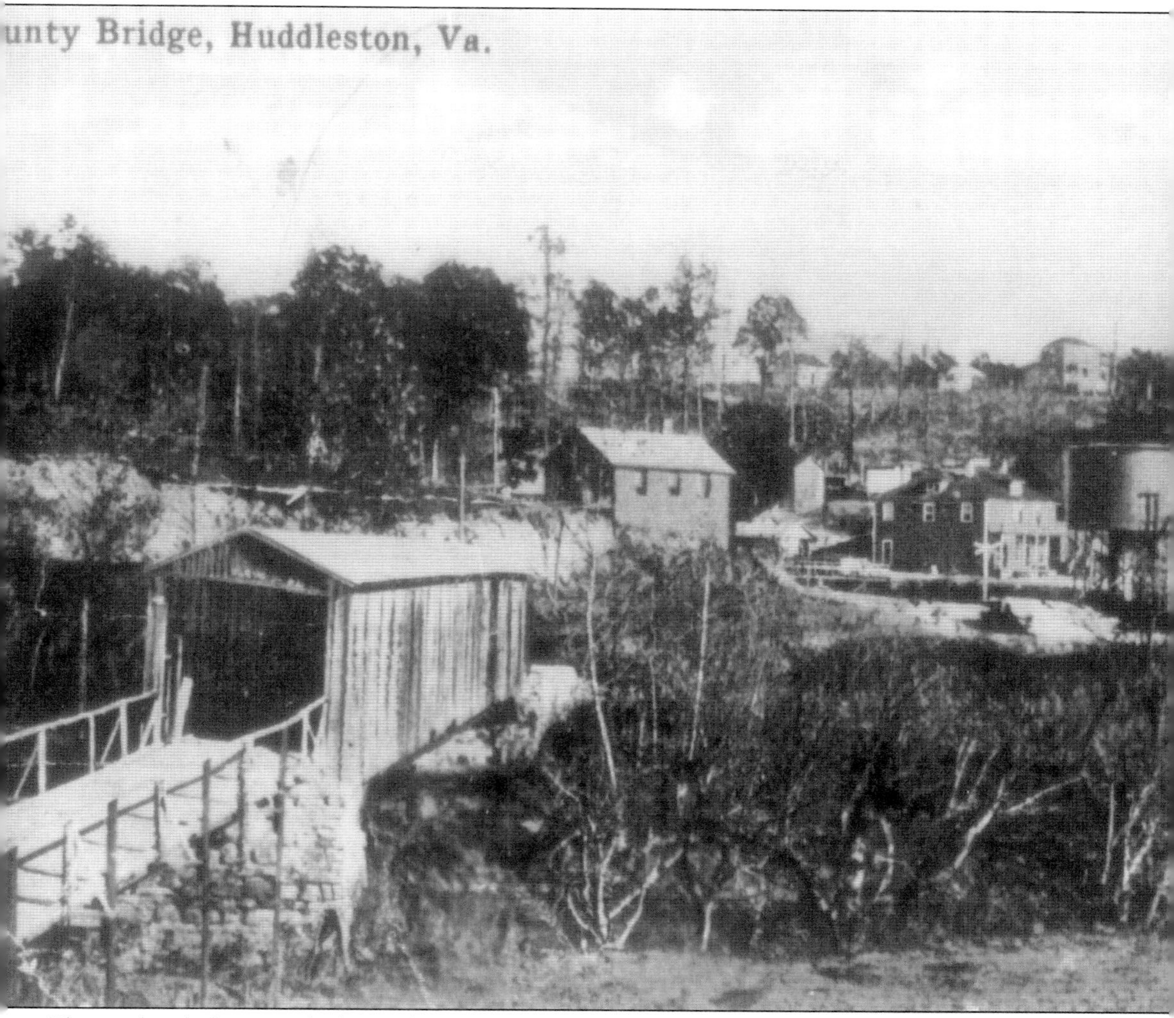

This undated photograph shows the village of Huddleston in Bedford County during the early 1920s. Note the location of the covered bridge over the river before looking at the next picture. (Courtesy of Frank Ewald.)

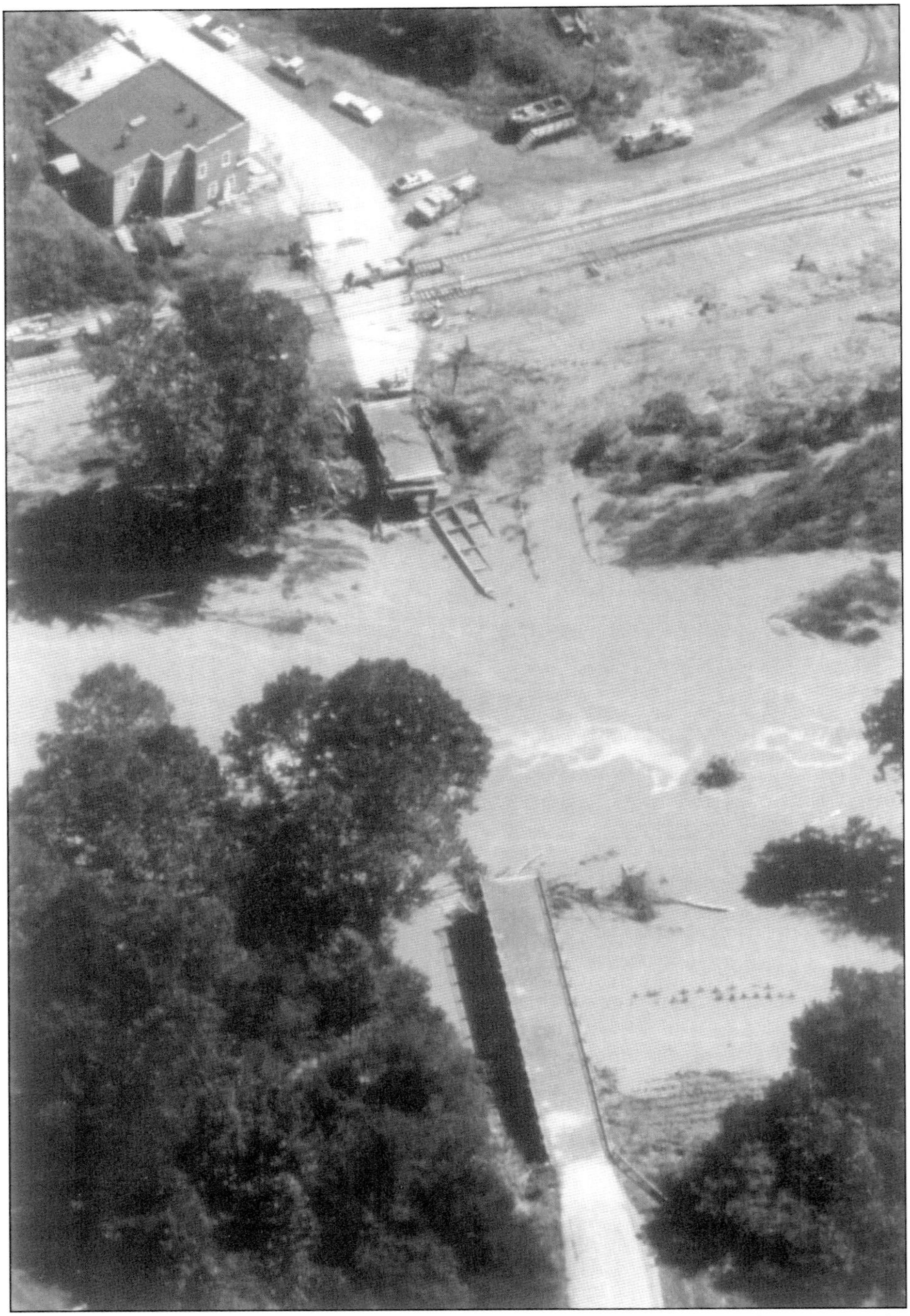

This aerial view of Huddleston was taken after the 1985 flood. According to the U.S. Geological Survey for the water resources of Virginia, "Heavy rainfall from October 31 through November 6, 1985, caused record-breaking floods over a large region, including western and northern Virginia. Most of the rain fell on November 4 and 5 and was indirectly related to Hurricane Juan. New maximum peak discharges were recorded at 63 stream gauging stations during this flood." In the Roanoke area, 10 people died as a result of this flood, 22 throughout Virginia. (Courtesy of Frank Ewald.)

Two

DOWN ON THE FARM

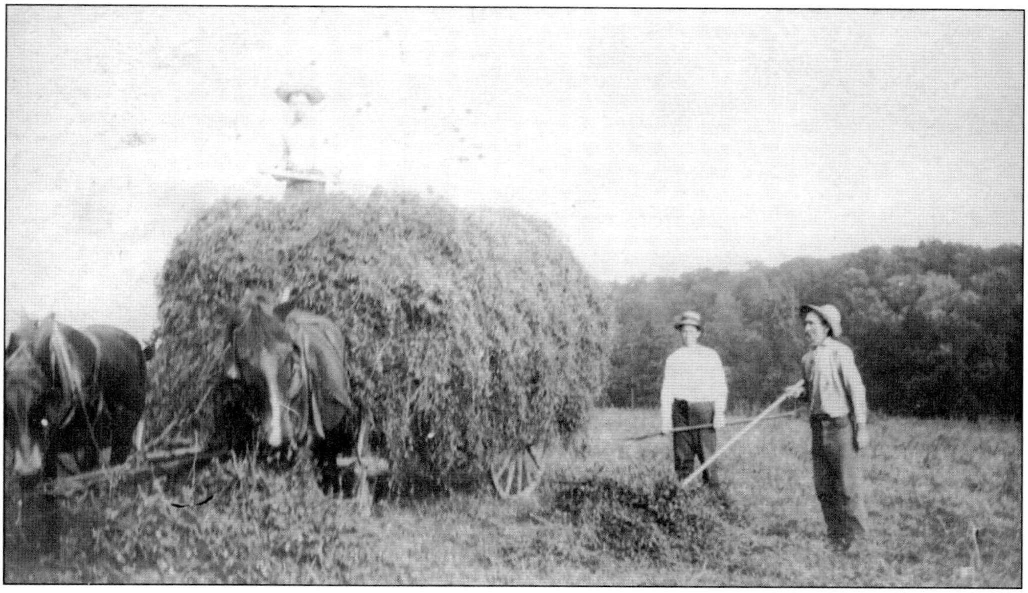

These farm boys are showing that necessity is the mother of invention. Before the emergence of square hay-balers and their successor, round balers, the pitchfork and elbow grease were the dominant technology in 1920s Bedford County hayfields. (Courtesy of Sam and Debbie Rucker.)

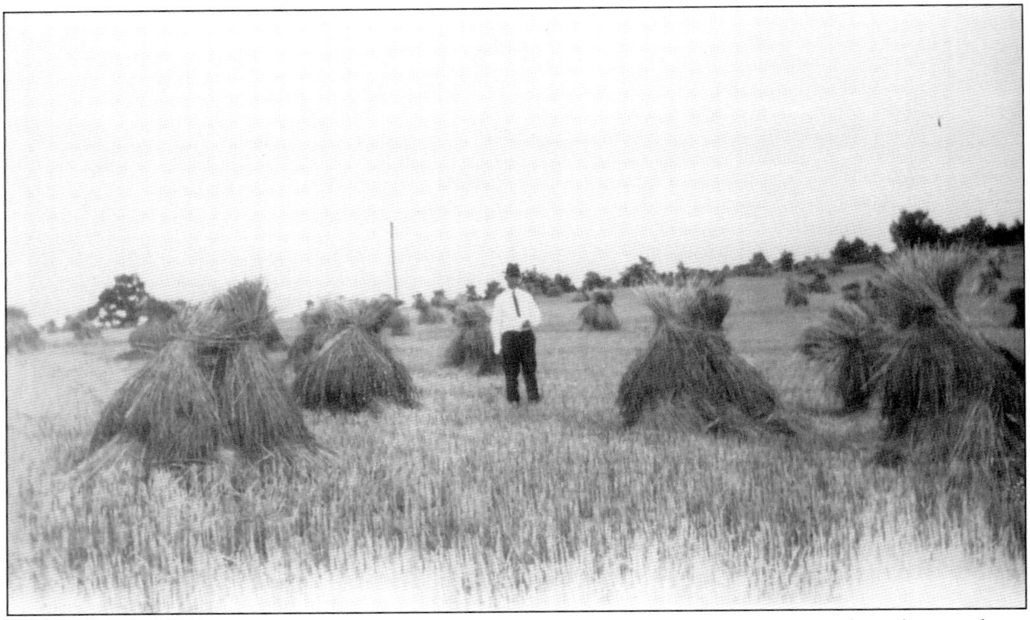

Cutting wheat by hand and scythe could make any farmer or, reaper, grim. Before the combine harvester, the wheat stalk was cut and then bound into shocks before being transported from the field. Here Key Phelps is standing in a field of wheat shocks on his farm in Bedford County. (Courtesy of Dr. James Rucker Jr.)

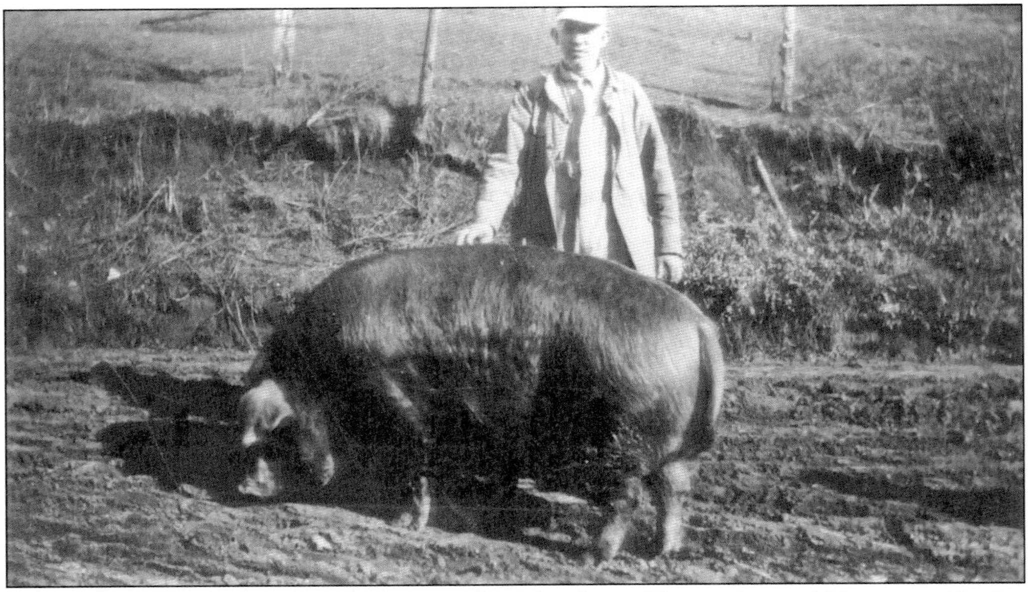

The author's father once remarked that a farm family could consider itself fortunate if, after raising hogs for slaughter, they could keep the hams for their personal consumption instead of having to sell them for extra income. Mary Rucker, sister of Dr. Sam Rucker Jr., took this photograph on December 9, 1928, and captioned it, "Mr. Forbes and 'The Pig'." Pork, including all parts except for "the squeal," was an essential part of the diet of farm families during this era. Judging from the size of this animal, Forbes was able to live "high on the hog." (Courtesy Dr. James Rucker Jr.)

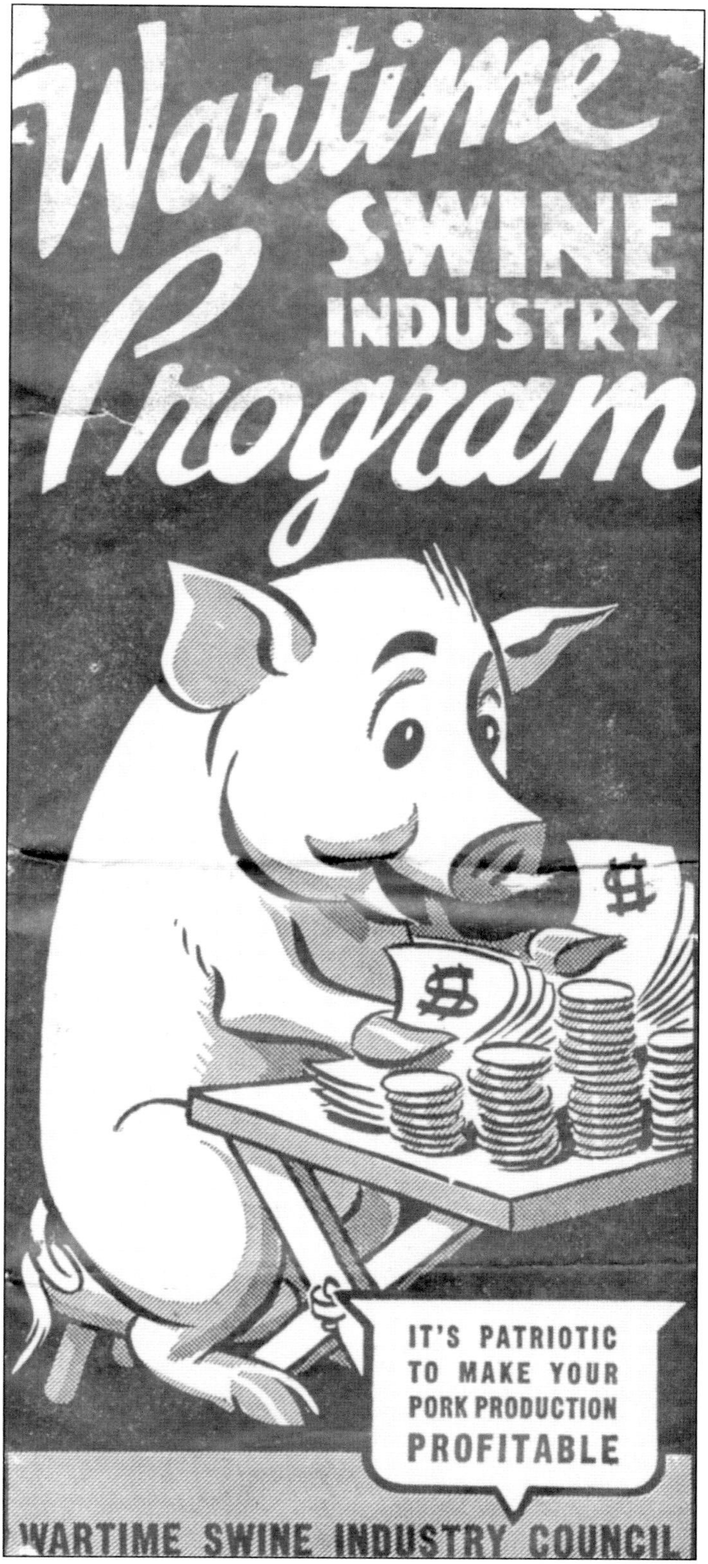

This Wartime Swine Industry Council pamphlet was distributed to producers "because of the current shortage in the livestock feed supply of the Nation." Inside, the preamble reads, "Meetings were called in Chicago by the Feed and Livestock Branch of the War Food Administration to present the problems immediately facing the swine industry with respect to present hog production levels and the feed shortage." Hog producers were called on to "reach the production goal set by the War Food Administration of a reduction of 11 percent in the fall pig crop of 1944, below the fall pig crop of 1943." Feed is wasted and profits are lost, according to the pamphlet, by "scrub breeding stock, sows that raise small litters, pigs that are squealers at birth and runts at weaning time, inadequate housing and shade, pigs that are not healthy and slip shod feeding methods." (Courtesy of the author.)

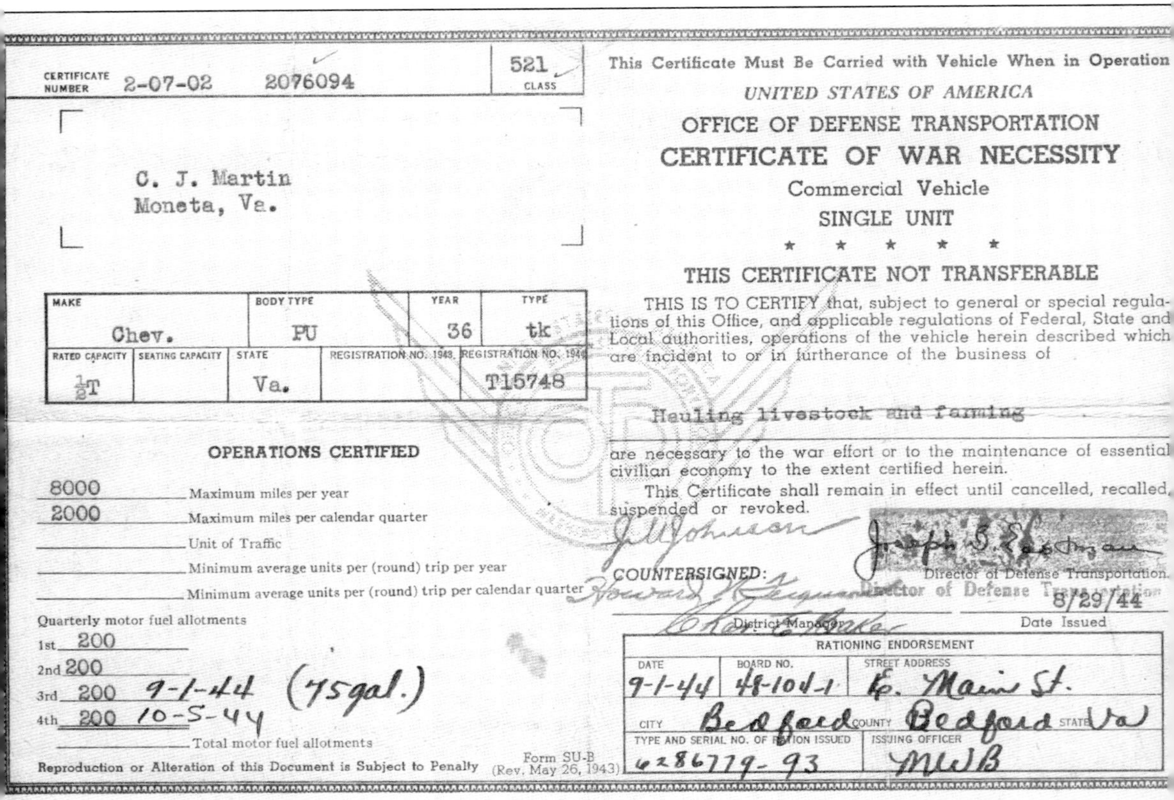
CERTIFICATE NUMBER 2-07-02 2076094

521 CLASS

This Certificate Must Be Carried with Vehicle When in Operation

UNITED STATES OF AMERICA

OFFICE OF DEFENSE TRANSPORTATION

CERTIFICATE OF WAR NECESSITY

Commercial Vehicle

SINGLE UNIT

* * * * *

THIS CERTIFICATE NOT TRANSFERABLE

C. J. Martin
Moneta, Va.

MAKE	BODY TYPE	YEAR	TYPE
Chev.	PU	36	tk

RATED CAPACITY	SEATING CAPACITY	STATE	REGISTRATION NO. 1943	REGISTRATION NO. 1944
½T		Va.		T15748

THIS IS TO CERTIFY that, subject to general or special regulations of this Office, and applicable regulations of Federal, State and Local authorities, operations of the vehicle herein described which are incident to or in furtherance of the business of

Hauling livestock and farming

are necessary to the war effort or to the maintenance of essential civilian economy to the extent certified herein.

This Certificate shall remain in effect until cancelled, recalled, suspended or revoked.

J. M. Johnson

Joseph B. Eastman
Director of Defense Transportation.

COUNTERSIGNED:

Howard F. Ferguson Director of Defense Transportation

District Manager

8/29/44
Date Issued

OPERATIONS CERTIFIED

8000 Maximum miles per year
2000 Maximum miles per calendar quarter
Unit of Traffic
Minimum average units per (round) trip per year
Minimum average units per (round) trip per calendar quarter

Quarterly motor fuel allotments
1st 200
2nd 200
3rd 200 9-1-44 (75 gal.)
4th 200 10-5-44
Total motor fuel allotments

RATIONING ENDORSEMENT

DATE	BOARD NO.	STREET ADDRESS
9-1-44	48-104-1	E. Main St.

CITY Bedford COUNTY Bedford STATE Va

TYPE AND SERIAL NO. OF RATION ISSUED	ISSUING OFFICER
6286779-93	MWB

Reproduction or Alteration of this Document is Subject to Penalty

Form SU-B
(Rev. May 26, 1943)

During World War II, food, clothing, gasoline, and other commodities were rationed for the war effort. Vehicles required for transportation and for farm use required documentation, as shown by this Certificate of War Necessity, Commercial Vehicle, Single Unit. Issued August 29, 1944, it lists the purposes for which C. J. Martin needed it: hauling livestock and farming. Note that operations for his 1936 Chevrolet pickup were certified for a maximum of 8,000 miles per year. (Courtesy of the author.)

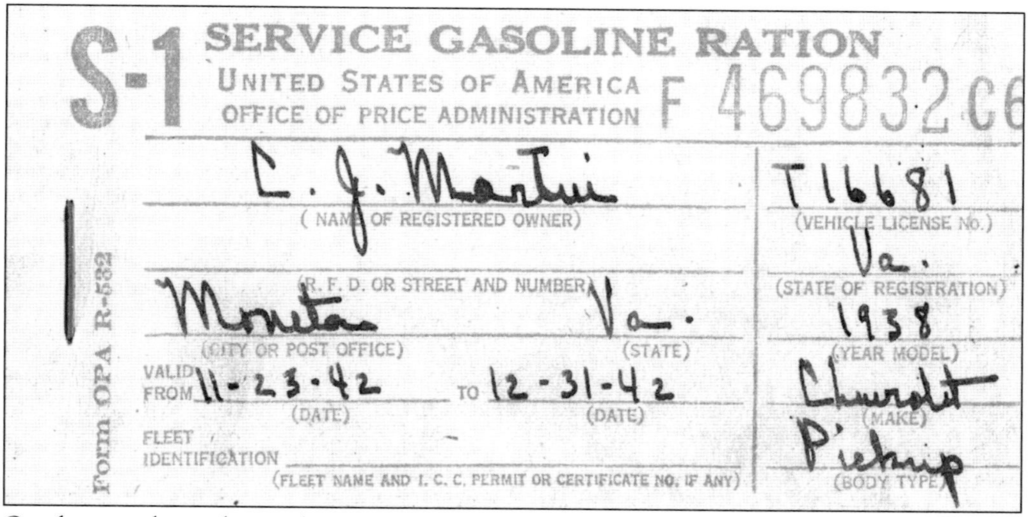

S-1 SERVICE GASOLINE RATION
UNITED STATES OF AMERICA
OFFICE OF PRICE ADMINISTRATION
F 469832 C6

Form OPA R-582

C. J. Martin
(NAME OF REGISTERED OWNER)

(R. F. D. OR STREET AND NUMBER)

Moneta Va.
(CITY OR POST OFFICE) (STATE)

VALID FROM 11-23-42 TO 12-31-42
(DATE) (DATE)

FLEET IDENTIFICATION
(FLEET NAME AND I. C. C. PERMIT OR CERTIFICATE NO. IF ANY)

T16681
(VEHICLE LICENSE NO.)

Va.
(STATE OF REGISTRATION)

1938
(YEAR MODEL)

Chevrolet
(MAKE)

Pickup
(BODY TYPE)

Gasoline purchases during the 1940s required stamps contained in booklets like the ones shown here. According to Wikipedia, "one had to appear before a local board to certify a need for gas and ownership of no more than five tires. All tires in excess of five per driver were confiscated by the government, because of rubber shortages. An A card was the lowest priority of gas rationing and entitled the holder to 3 to 4 gallons of gas per week. B cards were issued to workers in the military industry, entitling their holder up to 8 gallons of gas per week. C cards were granted to persons deemed very essential to the war effort, such as doctors. T rations were made available for truckers." (Courtesy of the author.)

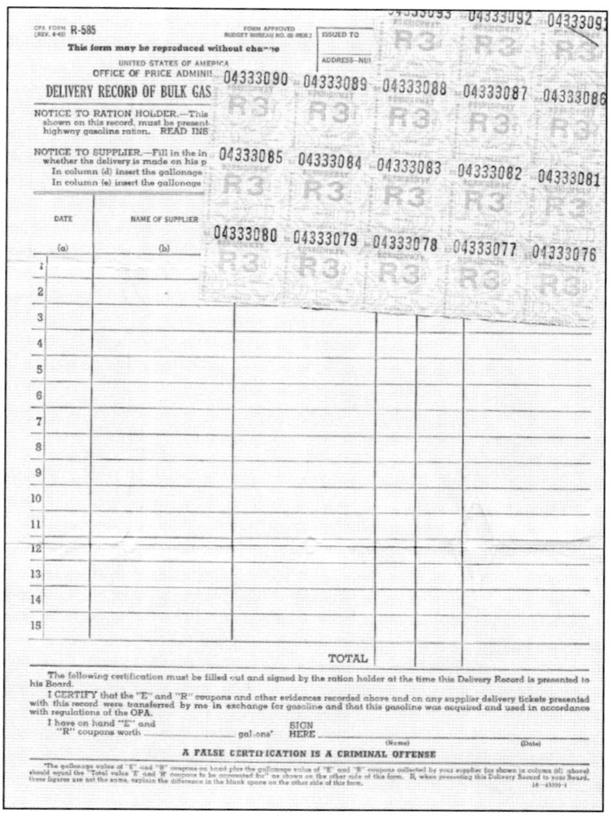

OPA FORM R-585 (REV. 4-43)
FORM APPROVED BUDGET BUREAU NO.
This form may be reproduced without change
ISSUED TO
ADDRESS—NU
UNITED STATES OF AMERICA
OFFICE OF PRICE ADMINIS
DELIVERY RECORD OF BULK GAS
NOTICE TO RATION HOLDER.—This shown on this record, must be present highway gasoline ration. READ INS
NOTICE TO SUPPLIER.—Fill in the in whether the delivery is made on his p
In column (d) insert the gallonage
In column (e) insert the gallonage

04333092 04333091
04333090 04333089 04333088 04333087 04333086
04333085 04333084 04333083 04333082 04333081
04333080 04333079 04333078 04333077 04333076
R3

	DATE (a)	NAME OF SUPPLIER (b)
1		
2		
3		
4		
5		
6		
7		
8		
9		
10		
11		
12		
13		
14		
15		
	TOTAL	

The following certification must be filled out and signed by the ration holder at the time this Delivery Record is presented to his Board.

I CERTIFY that the "E" and "R" coupons and other evidences recorded above and on any supplier delivery tickets presented with this record were transferred by me in exchange for gasoline and that this gasoline was acquired and used in accordance with regulations of the OPA.

I have on hand "E" and "R" coupons worth ______ gallons* SIGN HERE ______ (Name) (Date)

A FALSE CERTIFICATION IS A CRIMINAL OFFENSE

For the period from July 31, 1945, through January 31, 1946, "R" coupons representing 30 gallons of "non-highway" gasoline were issued to Bedford County farmer Carlton Martin by the War Price and Rationing Board. This notice appears on the document: "This delivery record, properly completed, and any delivery tickets covering deliveries not shown on this record must be presented to your War Price and Rationing Board with your next application or request for non-highway gasoline ration." The local board was in Bedford, Virginia. (Courtesy of the author.)

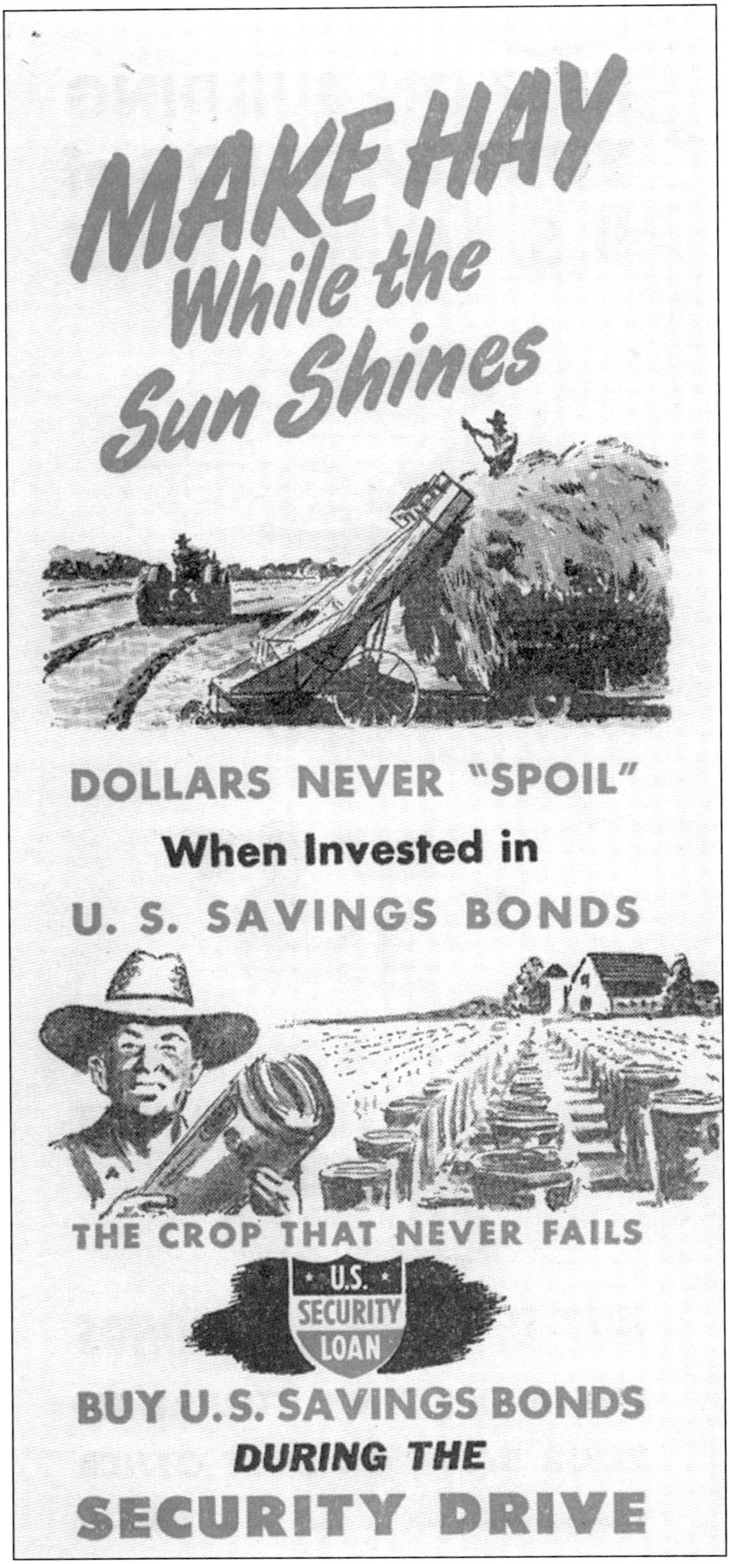

According to the Ad Council, from January 15, 1942, to August 14, 1945, advertising "focused on informing the American people what needed to be done to win World War II quickly and buying war bonds was near the top of that list. During that time, American advertising and media businesses contributed an estimated $350 million worth of space and time in support of war bond promotion and approximately 85 million Americans bought more than 800 million war bonds." After World War II, war bonds were called savings bonds. (Courtesy of the author.)

According to the U.S. Department of Agriculture, the Agricultural Adjustment Act of 1938 authorized quotas on an acreage allotment basis for various types of tobacco: "Quotas for dark air-cured, fire-cured, sun-cured, cigar filler & binder tobaccos are treated differently than burley and flue-cured tobaccos." This allotment for Harold Martin's Bedford County farm was signed November 16, 1948, for 4 acres for the period beginning October 1, 1949, and ending September 30, 1950. (Courtesy of the author.)

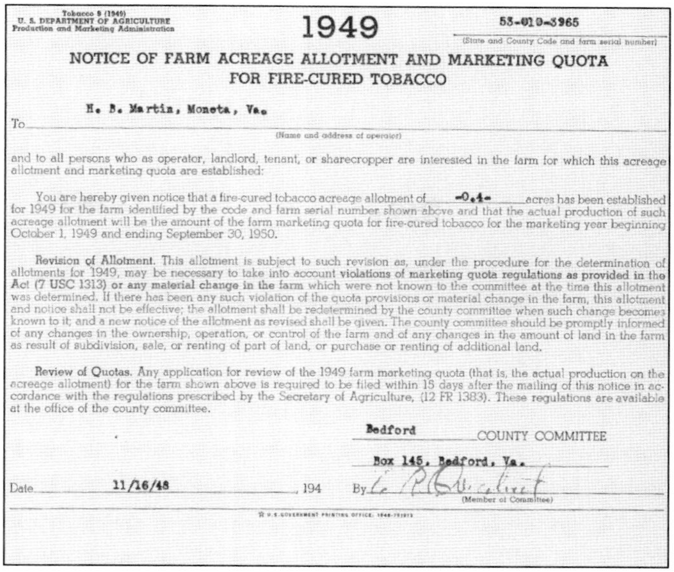

Tobacco 8 (1949)
U. S. DEPARTMENT OF AGRICULTURE
Production and Marketing Administration

1949

53-019-3965
(State and County Code and farm serial number)

NOTICE OF FARM ACREAGE ALLOTMENT AND MARKETING QUOTA FOR FIRE-CURED TOBACCO

To H. B. Martin, Moneta, Va.
(Name and address of operator)

and to all persons who as operator, landlord, tenant, or sharecropper are interested in the farm for which this acreage allotment and marketing quota are established:

You are hereby given notice that a fire-cured tobacco acreage allotment of -0.4- acres has been established for 1949 for the farm identified by the code and farm serial number shown above and that the actual production of such acreage allotment will be the amount of the farm marketing quota for fire-cured tobacco for the marketing year beginning October 1, 1949 and ending September 30, 1950.

Revision of Allotment. This allotment is subject to such revision as, under the procedure for the determination of allotments for 1949, may be necessary to take into account **violations of marketing quota regulations as provided in the Act (7 USC 1313) or any material change in the farm** which were not known to the committee at the time this allotment was determined. If there has been any such violation of the quota provisions or material change in the farm, this allotment and notice shall not be effective; the allotment shall be redetermined by the county committee when such change becomes known to it; and a new notice of the allotment as revised shall be given. The county committee should be promptly informed of any changes in the ownership, operation, or control of the farm and of any changes in the amount of land in the farm as result of subdivision, sale, or renting of part of land, or purchase or renting of additional land.

Review of Quotas. Any application for review of the 1949 farm marketing quota (that is, the actual production on the acreage allotment) for the farm shown above is required to be filed within 15 days after the mailing of this notice in accordance with the regulations prescribed by the Secretary of Agriculture, (12 FR 1383). These regulations are available at the office of the county committee.

Bedford COUNTY COMMITTEE

Box 145, Bedford, Va.

Date 11/16/48, 194 By
(Member of Committee)

U.S. GOVERNMENT PRINTING OFFICE: 1948-793973

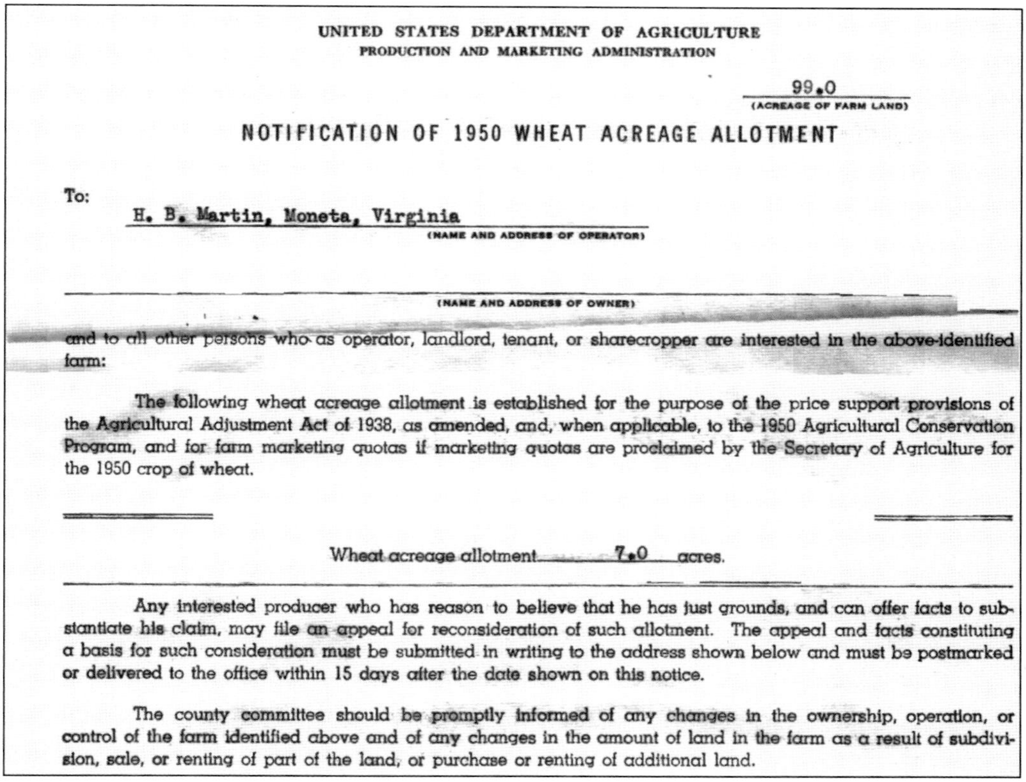

UNITED STATES DEPARTMENT OF AGRICULTURE
PRODUCTION AND MARKETING ADMINISTRATION

99.0
(ACREAGE OF FARM LAND)

NOTIFICATION OF 1950 WHEAT ACREAGE ALLOTMENT

To:
H. B. Martin, Moneta, Virginia
(NAME AND ADDRESS OF OPERATOR)

(NAME AND ADDRESS OF OWNER)

and to all other persons who as operator, landlord, tenant, or sharecropper are interested in the above-identified farm:

The following wheat acreage allotment is established for the purpose of the price support provisions of the Agricultural Adjustment Act of 1938, as amended, and, when applicable, to the 1950 Agricultural Conservation Program, and for farm marketing quotas if marketing quotas are proclaimed by the Secretary of Agriculture for the 1950 crop of wheat.

Wheat acreage allotment 7.0 acres.

Any interested producer who has reason to believe that he has just grounds, and can offer facts to substantiate his claim, may file an appeal for reconsideration of such allotment. The appeal and facts constituting a basis for such consideration must be submitted in writing to the address shown below and must be postmarked or delivered to the office within 15 days after the date shown on this notice.

The county committee should be promptly informed of any changes in the ownership, operation, or control of the farm identified above and of any changes in the amount of land in the farm as a result of subdivision, sale, or renting of part of the land, or purchase or renting of additional land.

Signed September 20, 1949, this wheat acreage allotment for 7 acres, which was in accordance with the Agricultural Adjustment Act of 1938, was given for 1950. Taken from the U.S. Code online, Title 7 reads in part, "It is declared to be the policy of Congress to continue the Soil Conservation and Domestic Allotment Act, as amended [16 U.S.C. 590a et seq.], for the purpose of conserving national resources, preventing the wasteful use of soil fertility, and of preserving, maintaining, and rebuilding the farm and ranch land resources in the national public interest." (Courtesy of the author.)

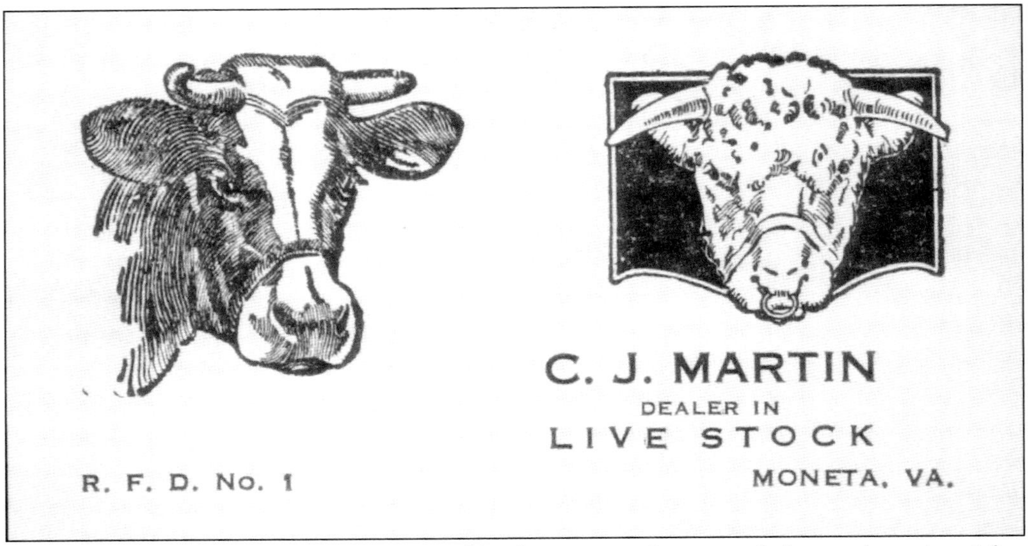

Carlton Martin believed in advertising, and his business card was simple but to the point. After the American Viscose Company in Roanoke closed in the 1950s, Martin, who had commuted there daily from Moneta, Virginia, became a full-time livestock dealer. In front of his home on White House Road in Moneta, a metal sign hung in the front yard that read, "I DOZE BUT NEVER CLOSE." This slogan was recommended to him by his friend, Clint Craighead, who noticed the business slogan on a sign in Virginia Beach. (Courtesy of the author.)

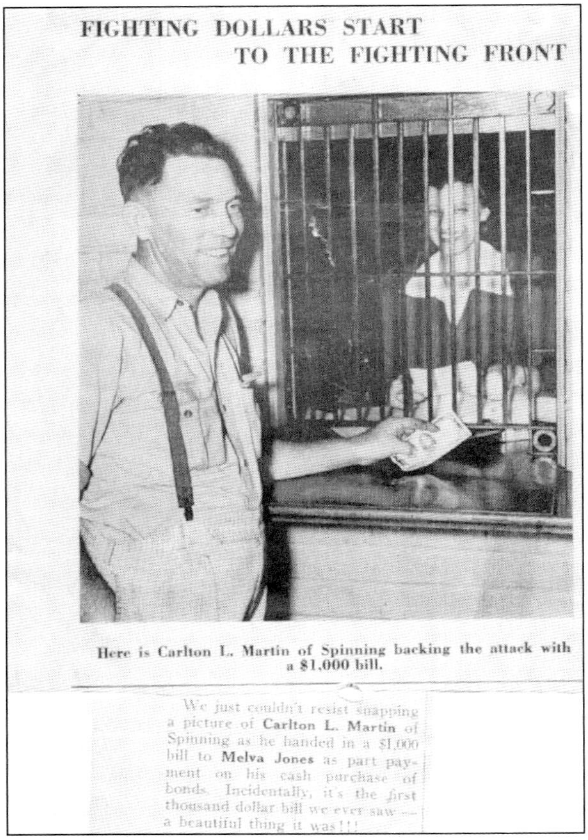
FIGHTING DOLLARS START TO THE FIGHTING FRONT

Here is Carlton L. Martin of Spinning backing the attack with a $1,000 bill.

We just couldn't resist snapping a picture of **Carlton L. Martin** of Spinning as he handed in a $1,000 bill to **Melva Jones** as part payment on his cash purchase of bonds. Incidentally, it's the first thousand dollar bill we ever saw — a beautiful thing it was!!!

Carlton Martin, who worked in the Spinning Department of the American Viscose Company, hands over a $1,000 bill to Viscose teller Melva Jones to buy war bonds. The event was unusual because, as the text of the article states, seeing $1,000 was not an everyday occasion, particularly for a Bedford County farmer. (Courtesy of the author.)

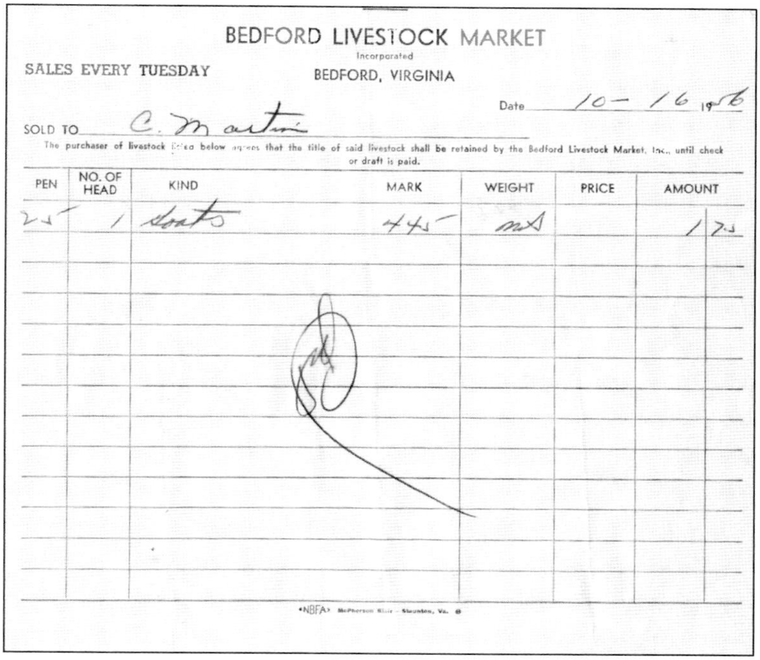

BEDFORD LIVESTOCK MARKET
Incorporated

SALES EVERY TUESDAY

BEDFORD, VIRGINIA

Date 10-16 1956

SOLD TO C. Martin

The purchaser of livestock listed below agrees that the title of said livestock shall be retained by the Bedford Livestock Market, Inc., until check or draft is paid.

PEN	NO. OF HEAD	KIND	MARK	WEIGHT	PRICE	AMOUNT
25	1	Goats	445	[illegible]		1 75

On October 16, 1956, Carlton Martin paid $1.75 for a goat at the Bedford Livestock Market. The receipt does not specify the type. However, as of May 2008, a commercial Boer nanny fetches $130 each. At www.GoatDealer.com, you can order the type of goat you want, pay with a credit card, and have it shipped to you for $2 per loaded mile. (Courtesy of the author.)

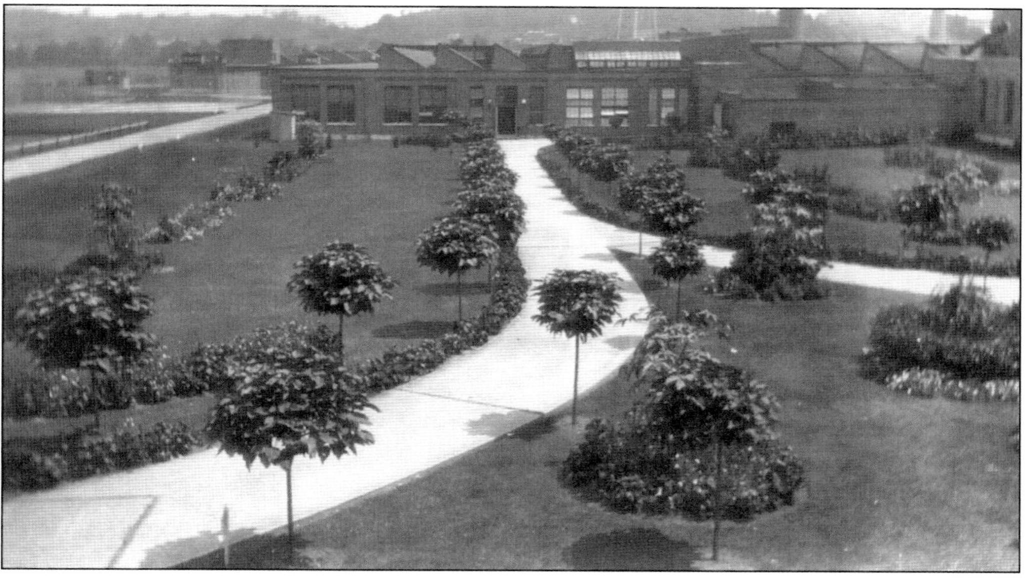

Until 1958, the American Viscose Company in Southeast Roanoke, Virginia, which opened in October 1917 for the production of the synthetic fabric rayon, provided an employment opportunity not only for Roanokers, but also for some Bedford County farmers who were able to make the daily commute. When the plant closed, 5,000 jobs were lost. (Courtesy of the author.)

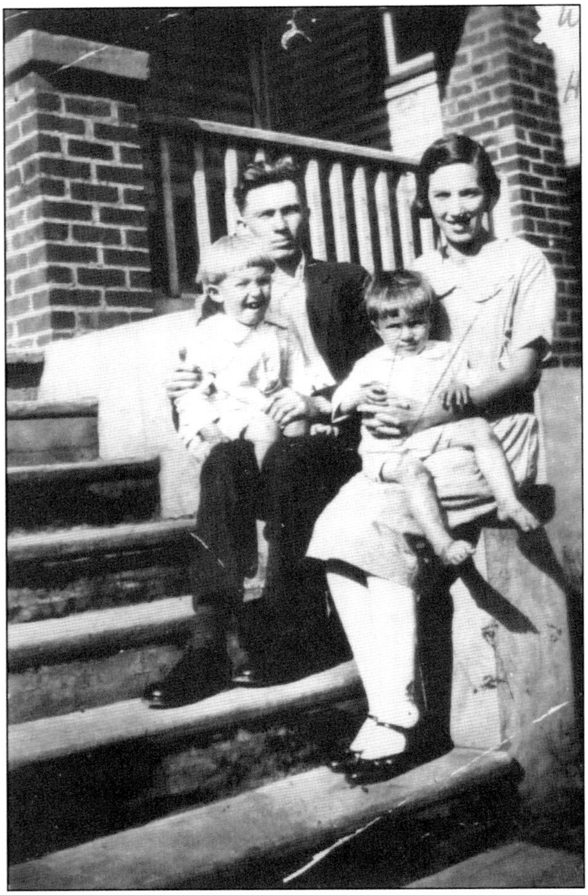

Because of winding country roads and less-comfortable automobiles than modern-day motorists enjoy, a daily commute from Bedford County to Southeast Roanoke to the Viscose plant was more of a challenge in the 1930s than it would be today. Pictured here are the author's grandparents, C. J. and Fawnie Martin, and two of their children, Willard (left) and Harold (right), when they lived for a period during the late 1920s on Montrose Avenue in Southeast Roanoke. (Courtesy of the author.)

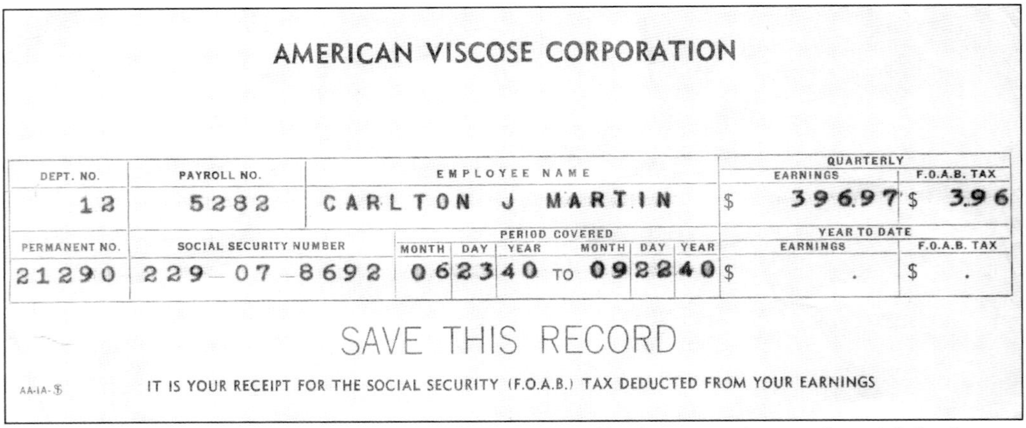

AMERICAN VISCOSE CORPORATION

DEPT. NO.	PAYROLL NO.	EMPLOYEE NAME	QUARTERLY EARNINGS	QUARTERLY F.O.A.B. TAX
12	5282	CARLTON J MARTIN	$ 396.97	$ 3.96

PERMANENT NO.	SOCIAL SECURITY NUMBER	PERIOD COVERED (MONTH DAY YEAR)	YEAR TO DATE EARNINGS	YEAR TO DATE F.O.A.B. TAX
21290	229 07 8692	062340 TO 092240	$.	$.

SAVE THIS RECORD

AA-IA-5

IT IS YOUR RECEIPT FOR THE SOCIAL SECURITY (F.O.A.B.) TAX DEDUCTED FROM YOUR EARNINGS

According to the audio journal of Harold Martin, a Bedford County farmer, farm life had many advantages: "We had plenty of food. We raised wheat and corn and pork. A fellow asked my brother, Willard, one time how we made it through the Depression. Willard said that we didn't even know it was going on because we had this farm and plenty to eat and our daddy had a job at the American Viscose in Roanoke. That lasted until the late 1950s when the plant closed down." This receipt for Social Security deductions from the paycheck of the author's grandfather also shows his earnings from June 23, 1940, to September 22, 1940, to be $396.97. (Courtesy of the author.)

This Christmas 1954 edition of the American Viscose Company newsletter, the *Avisco News*, was one of the last holiday issues produced. This issue was chock-full of Christmas-themed pictures and games for holiday-anxious company children to color and enjoy. Several years later, their parents would, hopefully, be employed elsewhere. (Courtesy of the author.)

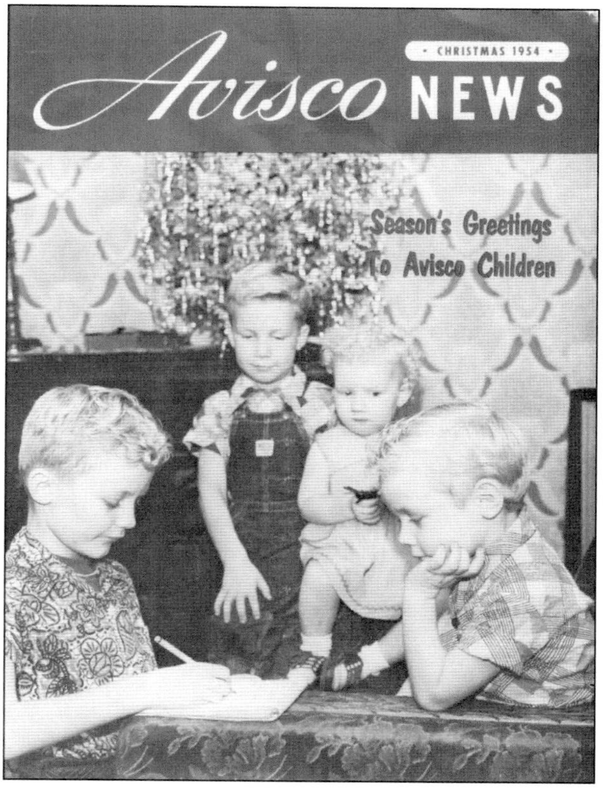

Wayne Turner of Bedford hooks up to a manure spreader in a contest in this 1950s photograph. (AGR0151, courtesy of the Virginia Cooperative Extension and the Digital Library and Archives, University Libraries, Virginia Polytechnic Institute and State University.)

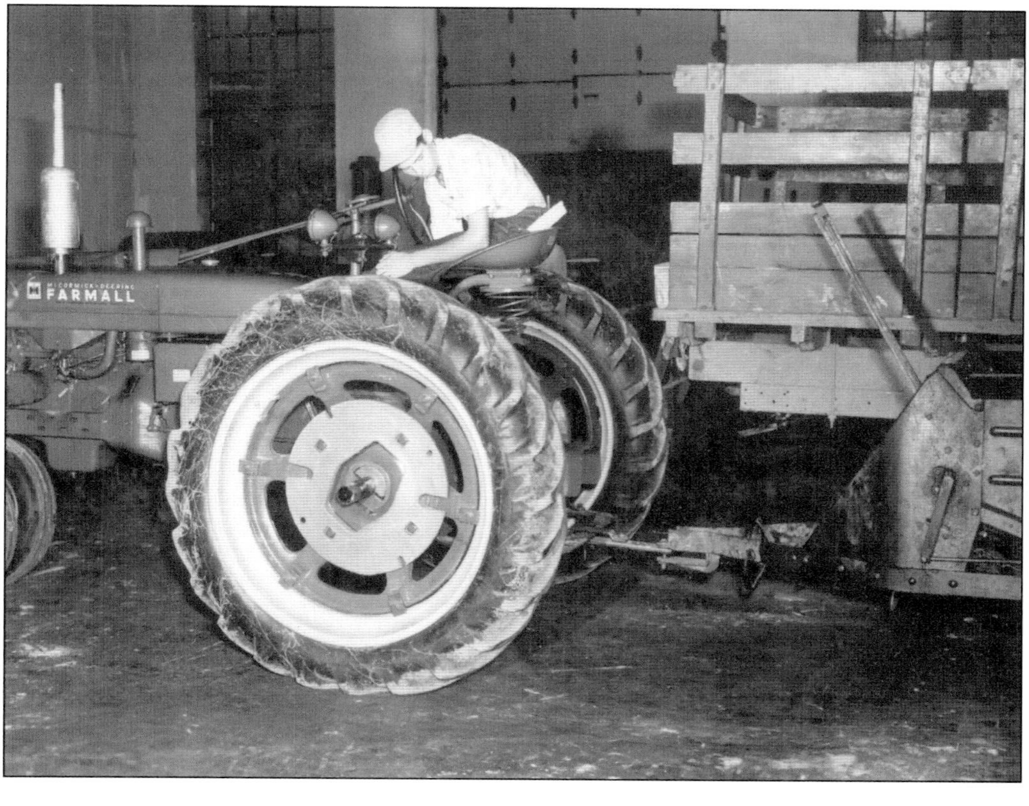

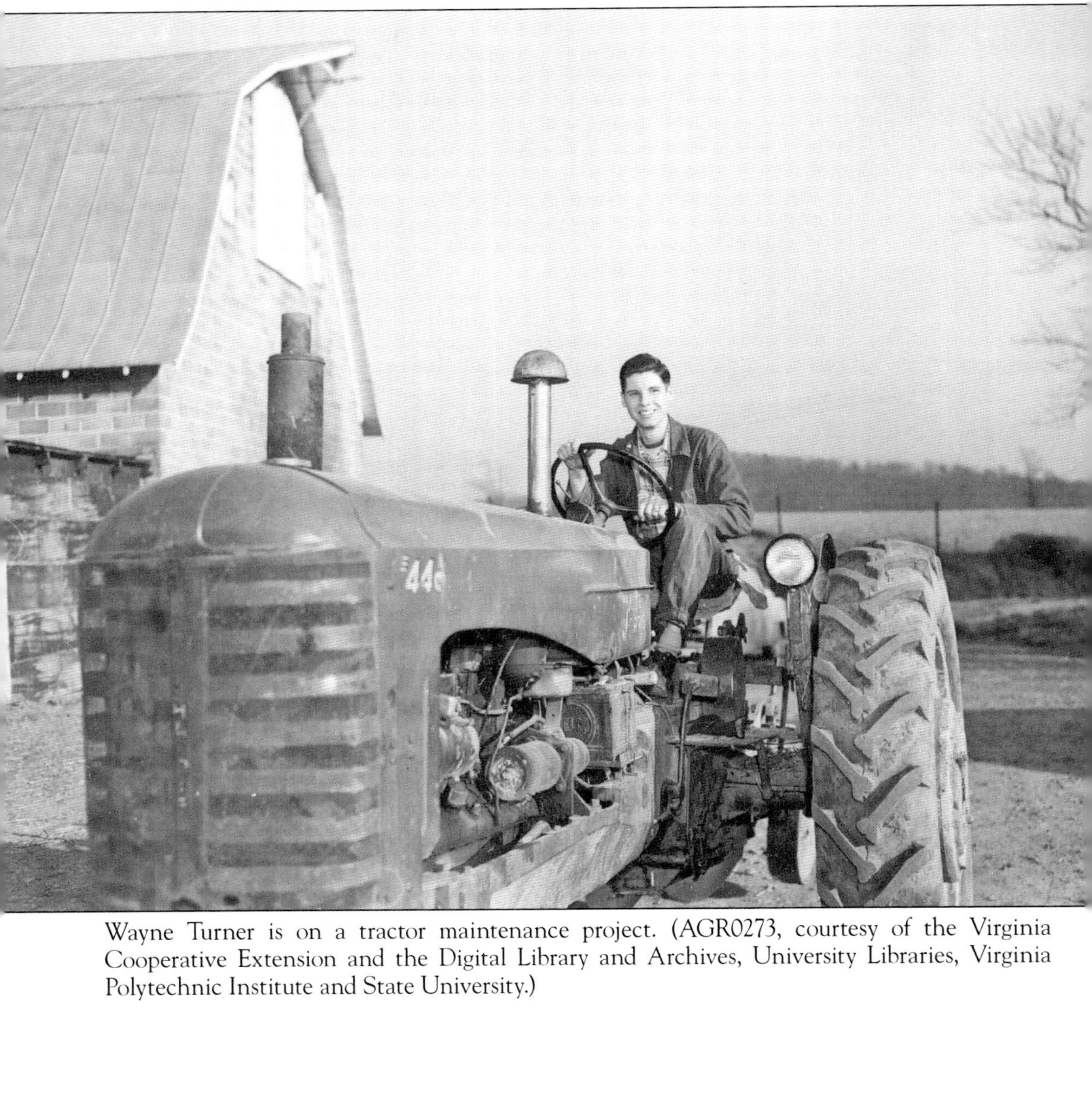

Wayne Turner is on a tractor maintenance project. (AGR0273, courtesy of the Virginia Cooperative Extension and the Digital Library and Archives, University Libraries, Virginia Polytechnic Institute and State University.)

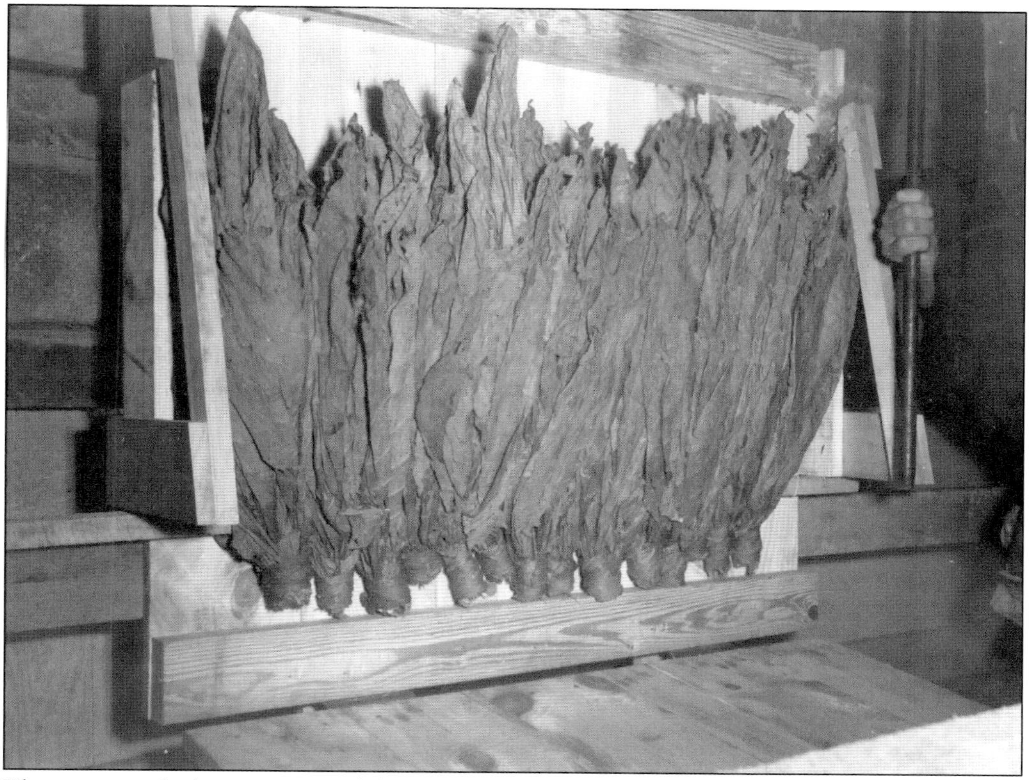

This is a typical tobacco press used during the days of Bedford County tobacco growing. (Courtesy of the Virginia Cooperative Extension and the Digital Library and Archives, University Libraries, Virginia Polytechnic Institute and State University.)

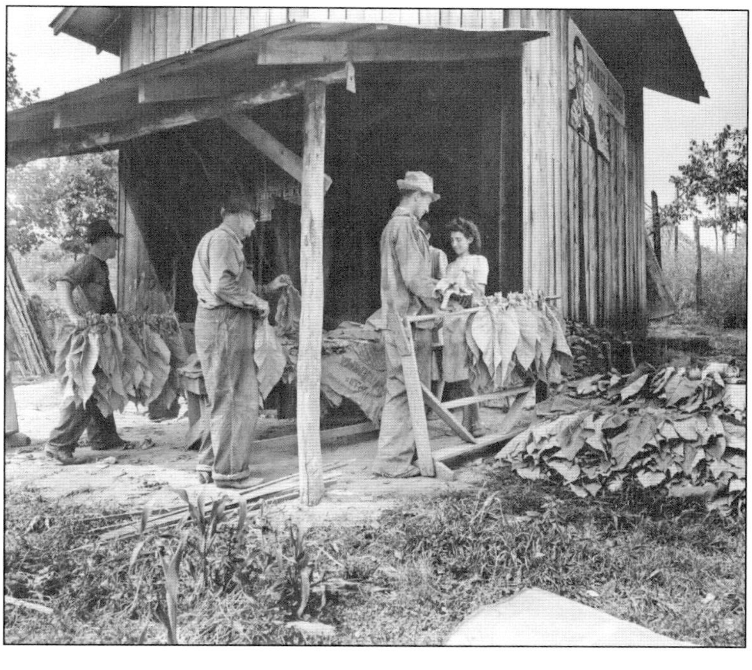

This would have been a common sight on Bedford County tobacco farms through the 1970s: workers manually processing tobacco leaves. Though processing leaves is not easy work, tobacco picking is a back-breaking labor. (NS4497, courtesy of the Norfolk and Western Historical Photograph Collection and the Digital Library and Archives, University Libraries, Virginia Polytechnic Institute and State University.)

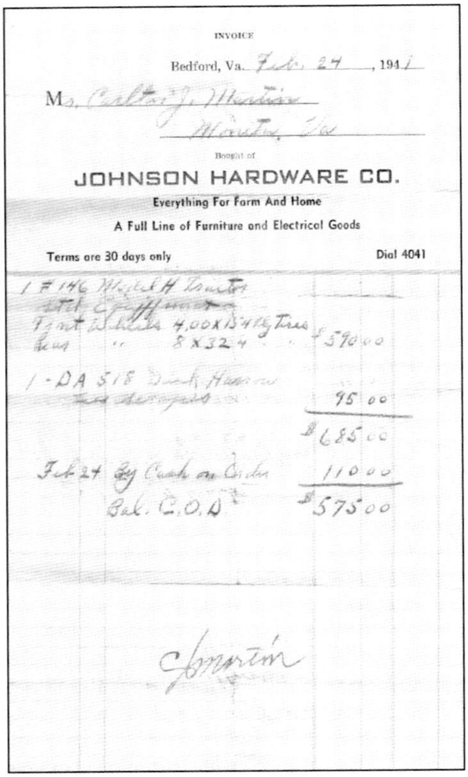
INVOICE

Bedford, Va. Feb. 24, 1941

Mr. Carlton J. Martin

Moneta, Va.

Bought of

JOHNSON HARDWARE CO.

Everything For Farm And Home

A Full Line of Furniture and Electrical Goods

Terms are 30 days only Dial 4041

1 #146 Model H Tractor

Front Wheels 4.00 x 15 4 Ply Tires

Rear " 8 x 32 4 $590.00

1-DA 518 Disk Harrow 95.00

$685.00

Feb 24 By Cash on Order 110.00

Bal. C.O.D. $575.00

C. J. Martin

This 1941 bill from the Johnson Hardware Store in Bedford, Virginia, is for a Model H tractor and a disc harrow. The price is $685, a price any Bedford County farmer in the 21st century would be glad to pay. (Courtesy of the author.)

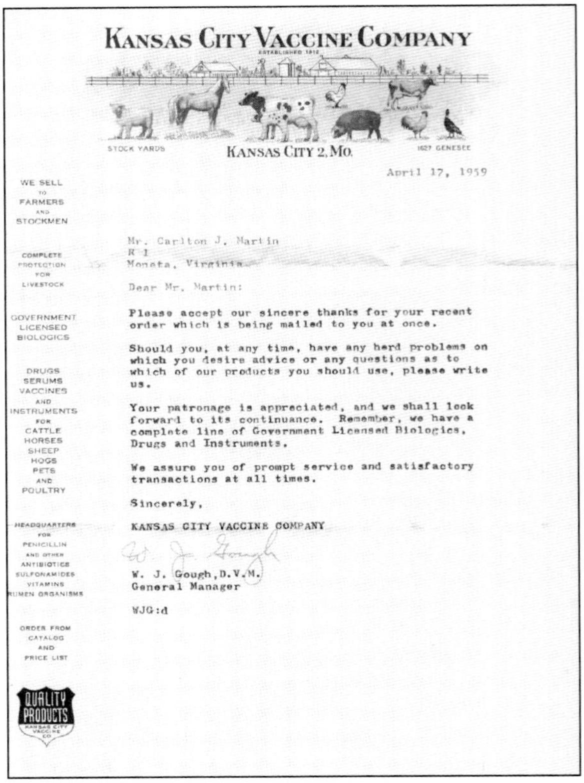
KANSAS CITY VACCINE COMPANY

ESTABLISHED 1912

STOCK YARDS KANSAS CITY 2, MO. 1627 GENESEE

April 17, 1959

WE SELL TO FARMERS AND STOCKMEN

COMPLETE PROTECTION FOR LIVESTOCK

GOVERNMENT LICENSED BIOLOGICS

DRUGS SERUMS VACCINES AND INSTRUMENTS FOR CATTLE HORSES SHEEP HOGS PETS AND POULTRY

HEADQUARTERS FOR PENICILLIN AND OTHER ANTIBIOTICS SULFONAMIDES VITAMINS RUMEN ORGANISMS

ORDER FROM CATALOG AND PRICE LIST

QUALITY PRODUCTS KANSAS CITY VACCINE CO.

Mr. Carlton J. Martin
R 1
Moneta, Virginia

Dear Mr. Martin:

Please accept our sincere thanks for your recent order which is being mailed to you at once.

Should you, at any time, have any herd problems on which you desire advice or any questions as to which of our products you should use, please write us.

Your patronage is appreciated, and we shall look forward to its continuance. Remember, we have a complete line of Government Licensed Biologics, Drugs and Instruments.

We assure you of prompt service and satisfactory transactions at all times.

Sincerely,

KANSAS CITY VACCINE COMPANY

W. J. Gough

W. J. Gough, D.V.M.
General Manager

WJG:d

This letter from the Kansas City Vaccine Company in 1959 referenced a payment of $10.75 by Carlton Martin for a hog holder and ear notcher. Some farm items were ordered by catalog well before the proliferation of catalog companies a half century later. (Courtesy of the author.)

If only automobile repairs were as cheap now as they were in 1950. This total bill from the Bud Speck garage in Huddleston is for $14.90, $5 of which was for the repair of a truck's master cylinder. (Courtesy of the author.)

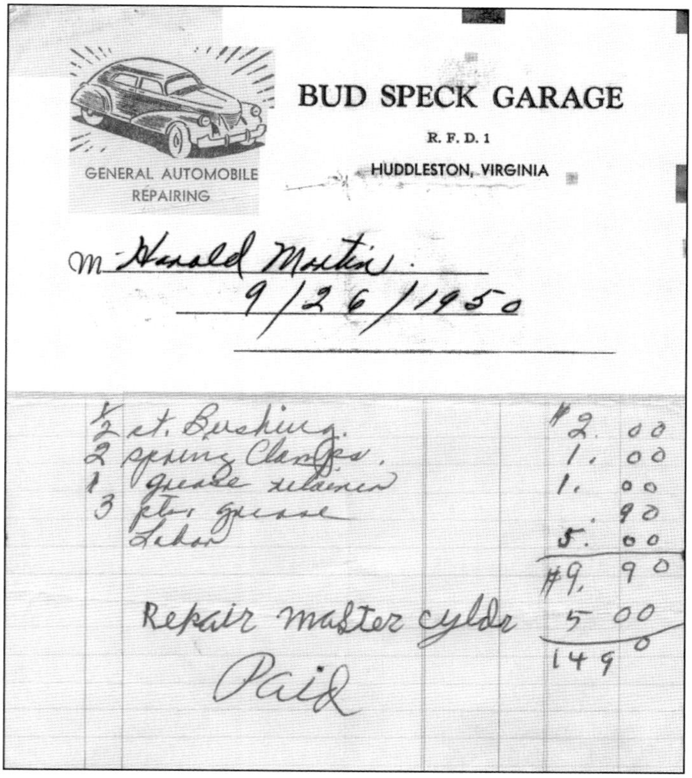

BUD SPECK GARAGE

R. F. D. 1

HUDDLESTON, VIRGINIA

GENERAL AUTOMOBILE REPAIRING

M Harold Martin

9/26/1950

½	st. Bushing	$2.00
2	spring Clamps	1.00
1	grease retainer	1.00
3	lbs. grease	.90
	Labor	5.00
		$9.90
	Repair master cyldr	5.00
		14.90

Paid

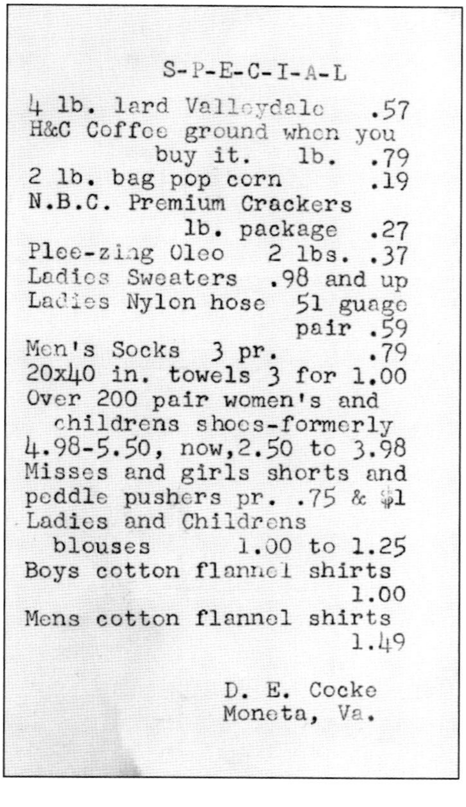

S-P-E-C-I-A-L

4 lb. lard Valleydale .57
H&C Coffee ground when you buy it. lb. .79
2 lb. bag pop corn .19
N.B.C. Premium Crackers lb. package .27
Plee-zing Oleo 2 lbs. .37
Ladies Sweaters .98 and up
Ladies Nylon hose 51 guage pair .59
Men's Socks 3 pr. .79
20x40 in. towels 3 for 1.00
Over 200 pair women's and childrens shoes-formerly 4.98-5.50, now,2.50 to 3.98
Misses and girls shorts and peddle pushers pr. .75 & $1
Ladies and Childrens blouses 1.00 to 1.25
Boys cotton flannel shirts 1.00
Mens cotton flannel shirts 1.49

D. E. Cocke
Moneta, Va.

This simple but effective postcard is a precursor to the glossy store advertisements that grocery store chains include in newspapers today. The D. E. Cocke General Store was located on State Route 122, across the road from the current location of the Shoprite Grocery Store. Note that H&C Coffee is a whopping 79¢ a pound. There is no postmark on the card. (Courtesy of the author.)

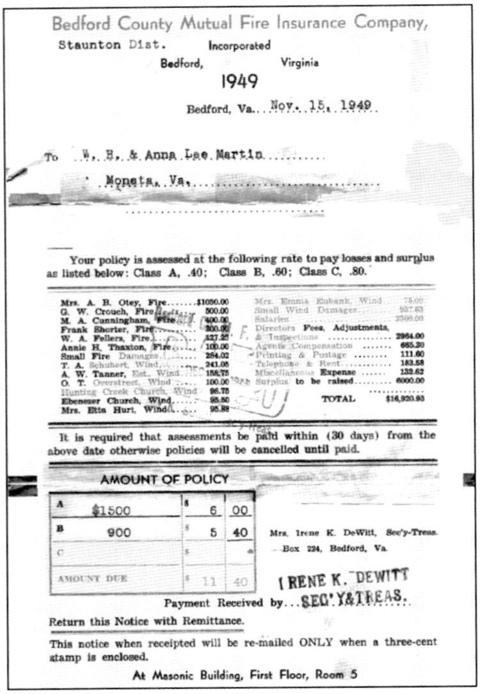

Bedford County Mutual Fire Insurance Company,
Staunton Dist. Incorporated
Bedford, Virginia
1949

Bedford, Va. Nov. 15, 1949

To W. H. & Anna Lee Martin
Moneta, Va.

Your policy is assessed at the following rate to pay losses and surplus as listed below: Class A, .40; Class B, .60; Class C, .80.

Mrs. A. B. Otey, Fire	$1050.00
G. W. Crouch, Fire	500.00
M. A. Cunningham, Fire	400.00
Frank Shorter, Fire	300.00
W. A. Fellers, Fire	127.25
Annie H. Thaxton, Fire	100.00
Small Fire Damages	284.02
T. A. Schubert, Wind	241.08
A. W. Tanner, Est. Wind	158.75
O. T. Overstreet, Wind	100.00
Hunting Creek Church, Wind	96.75
Ebenezer Church, Wind	95.80
Mrs. Etta Hurt, Wind	95.88
Mrs. Emma Eubank, Wind	75.00
Small Wind Damages	937.83
Salaries	2300.00
Directors Fees, Adjustments, & Inspections	2964.00
Agents Compensation	665.30
Printing & Postage	111.60
Telephone & Rent	183.58
Miscellaneous Expense	132.62
Surplus to be raised	6000.00
TOTAL	$16,920.93

It is required that assessments be paid within (30 days) from the above date otherwise policies will be cancelled until paid.

AMOUNT OF POLICY

		$	
A	$1500	6	00
B	900	5	40
C			
AMOUNT DUE		11	40

Mrs. Irene K. DeWitt, Sec'y-Treas.
Box 224, Bedford, Va.

Payment Received by... IRENE K. DEWITT SEC'Y&TREAS.

Return this Notice with Remittance.

This notice when receipted will be re-mailed ONLY when a three-cent stamp is enclosed.

At Masonic Building, First Floor, Room 5

For a brief period in the late 1940s, Willard and Anna Martin owned the John Kasey House on White House Road. This November 15, 1949, fire insurance bill is for $11.40. (Courtesy of the author.)

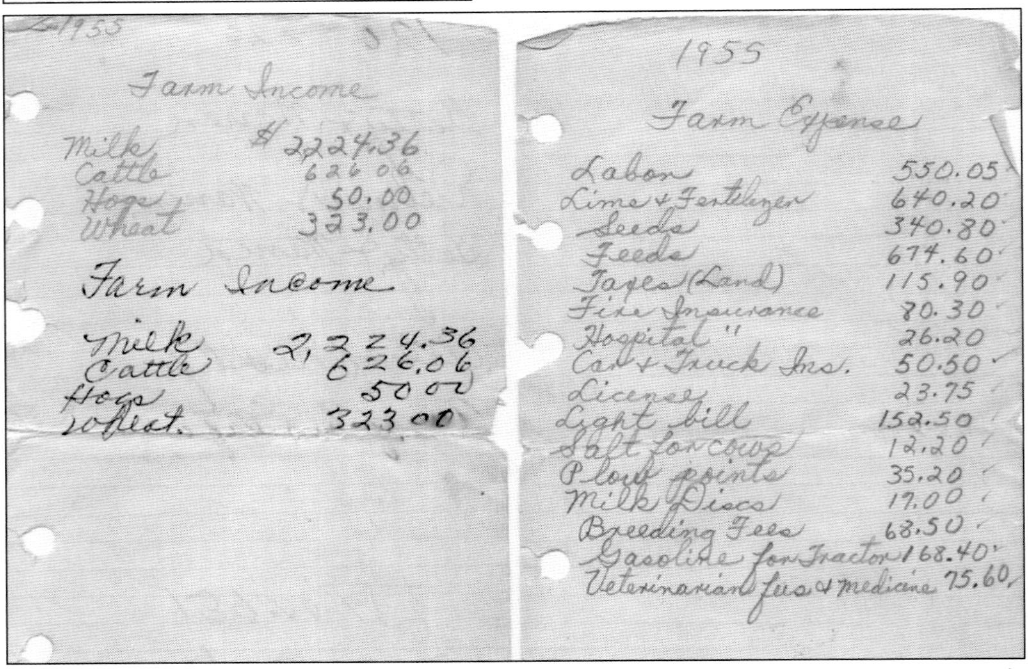

1955

Farm Income

Milk	$2224.36
Cattle	626.06
Hogs	50.00
Wheat	323.00

Farm Income

Milk	2,224.36
Cattle	626.06
Hogs	50.00
Wheat	323.00

1955

Farm Expense

Labor	550.05
Lime & Fertilizer	640.20
Seeds	340.80
Feeds	674.60
Taxes (Land)	115.90
Fire Insurance	80.30
Hospital "	26.20
Car & Truck Ins.	50.50
License	23.75
Light bill	152.50
Salt for cows	12.20
Plow points	35.20
Milk Discs	17.00
Breeding Fees	68.50
Gasoline for Tractor	168.40
Veterinarian fees & medicine	75.60

This simple but accurate income and expense summary shows a Bedford County farmer's profit snapshot for a portion of 1955. Harold Martin lists income from the production of milk, cattle, hogs, and wheat at $3,223.42. Expenses are $3,031.70 for the period. Though he was operating "in the black," these figures explain why many small farm owners did and, still do, need jobs off the farm in order to supplement their income. Milk production was once lucrative in Bedford County, but when the Coble Milk Transfer Station closed in Moneta, Martin was forced to seek employment off the farm. (Courtesy of the author.)

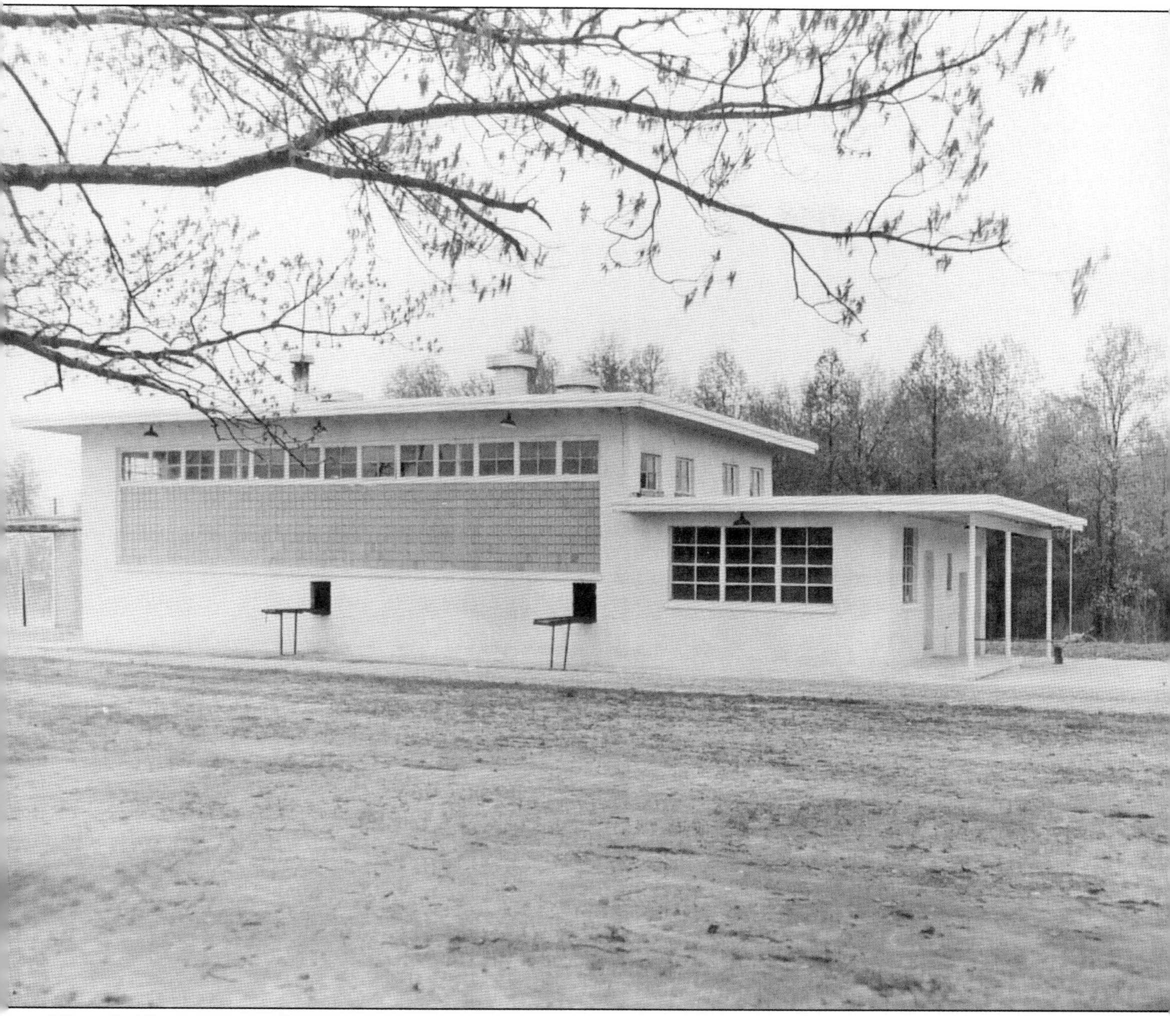

The Coble Dairy Milk Transfer Station in Moneta, which stood near the site of the current Lake Christian Ministries and is in close proximity to the Downtown Moneta restaurant and store complex, operated from the 1940s through the late 1960s. Farmers brought their milk in metal cans to the front of the facility at the two loading stations. Here it was placed in larger containers and then trucked to the South Boston facility, where it was manufactured into White House Brand cottage cheese. Other transfer stations were in Burkeville, Fork Union, and Stuart, Virginia. Finally Coble had a contract with the navy at Portsmouth, Virginia. Milk was frozen in half-gallon wax-coated containers for use by soldiers stationed on ships. (Courtesy of Reid Crouse.)

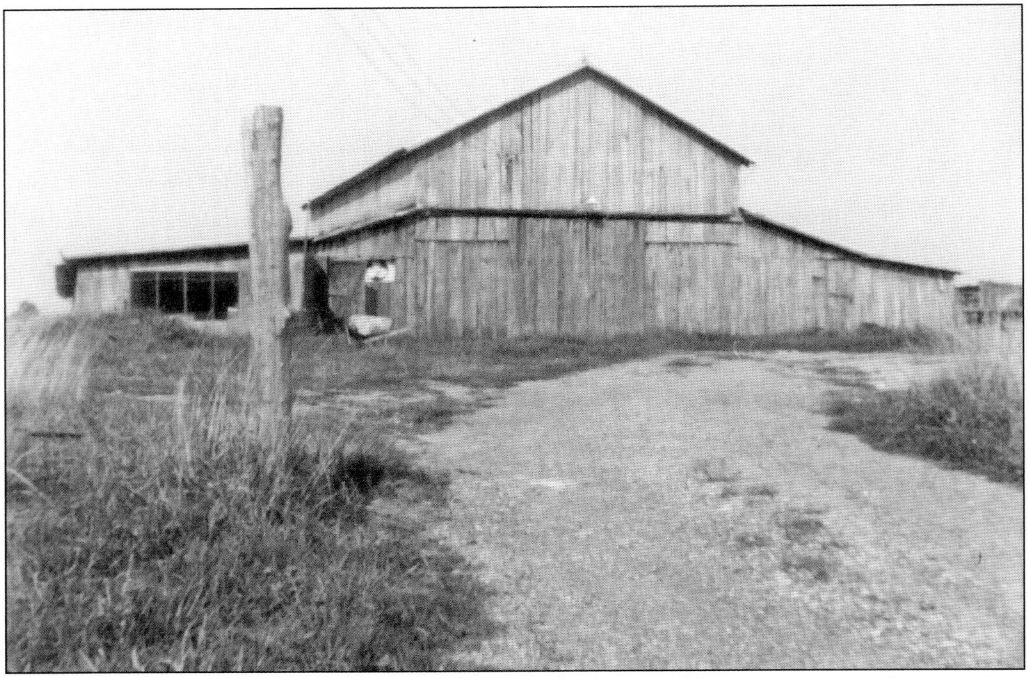

This is a typical dairy barn of the 1950s and 1960s. The large middle section was used to store hay for feed while the left side held holding areas for milking. In this model, the right side contained holding pens and feeding facilities to keep cattle "penned up" until they could be taken to market or, in some cases, for required vaccinations. The windowed left side of the structure was the area where freshly drawn milk would be poured into heavy metal cans and then lowered into a cooler before transfer to the Coble Dairy in Moneta. (Courtesy of the author.)

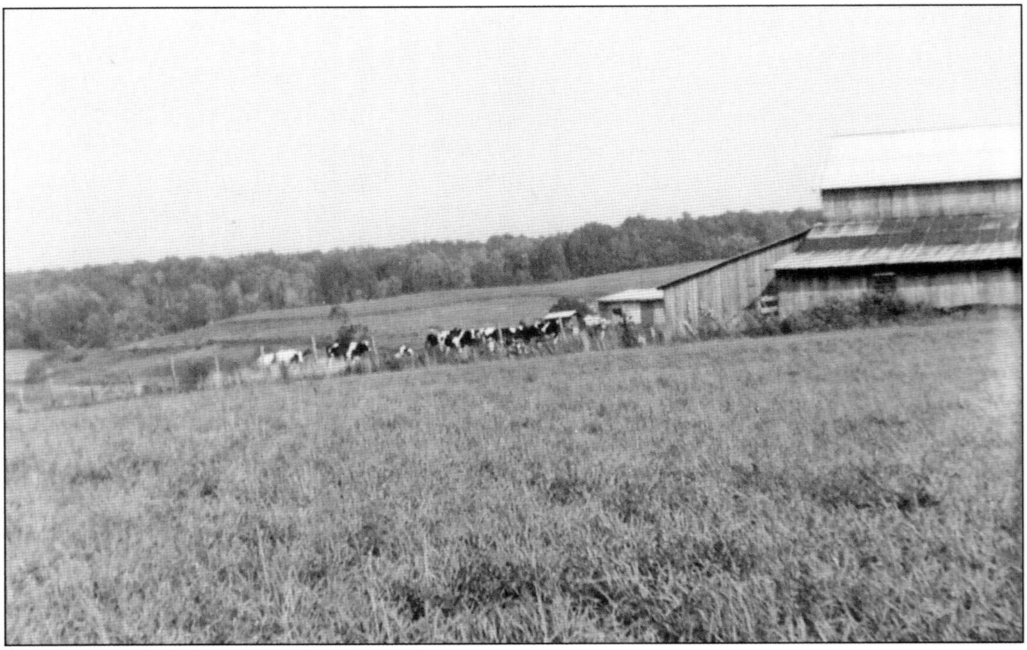

This side view of the same dairy barn shows milk cattle coming home for a feeding. The back section provided sheltered feeding facilities. (Courtesy of the author.)

This March 15, 1950, letter from Coble fieldman Frank Overstreet reminds producers that clean cans and utensils produce better milk for the customer. (Courtesy of the author.)

COBLE DAIRY PRODUCTS, INC.

MONETA, VIRGINIA

March 15, 1950

Dear Patron:

Now that spring is here, our thoughts naturally turn to spring cleaning - Whether it's the home or the farm.

When you start this spring job, let's not forget your milk cans and utensils.

Quality is being demanded by all of our customers.

Clean, sweet milk products are all they want.

Sour milk is a loss to everyone.

We are going to start our spring inspection soon and we are listing below what we should find:

Clean Cans - No milkstone inside and clean outside.

Clean Pails and Strainer - No milkstone inside and clean outside. (Dicoloid will do the job. Your hauler can get it for you).

B. K. used before every milking.

Dairy Cleanser and a hand brush used after each milking.

Cans and utensils stand bottoms-up in the open.

Of course, thoroughly cool each milking in a barrel or cooler.

COBLE DAIRY PRODUCTS

Frank Overstreet
Frank Overstreet
Fieldman

HAULING

EXCISE TAX

TOTAL

1996

TOTAL POUNDS

PRODUCER NUMBER 44 12

DATE	16	17	18	19	20	21	22	23	24	25	26	27	28	29	30	31
LBS.	126	142	140	144	133	102	144	123	105	133	140	126	120	117	105	105

DATE	POUNDS	TEST	BUTTER FAT	PRICE	GROSS	DEDUCTIONS HAULING	PRODUCERS ACCOUNT	CANS	NET AMOUNT
JAN 31 47	19.96	4.4	.878	4.69	93.61				93.61

A63040

SETTLEMENT STATEMENT

DETACH FROM CHECK AND RETAIN FOR YOUR RECORDS.
ANY ERRORS MUST BE REPORTED WITHIN 30 DAYS.

COBLE DAIRY PRODUCTS INC.

LEXINGTON, N. C.

This typical milk settlement statement card provided by Coble Dairy gives a test grade and total pounds produced for each pay period. The back of the card is a further summarization of the week's work and the most important figure of all: the pay, in this case, $74.20 for February 15, 1947. (Courtesy of the author.)

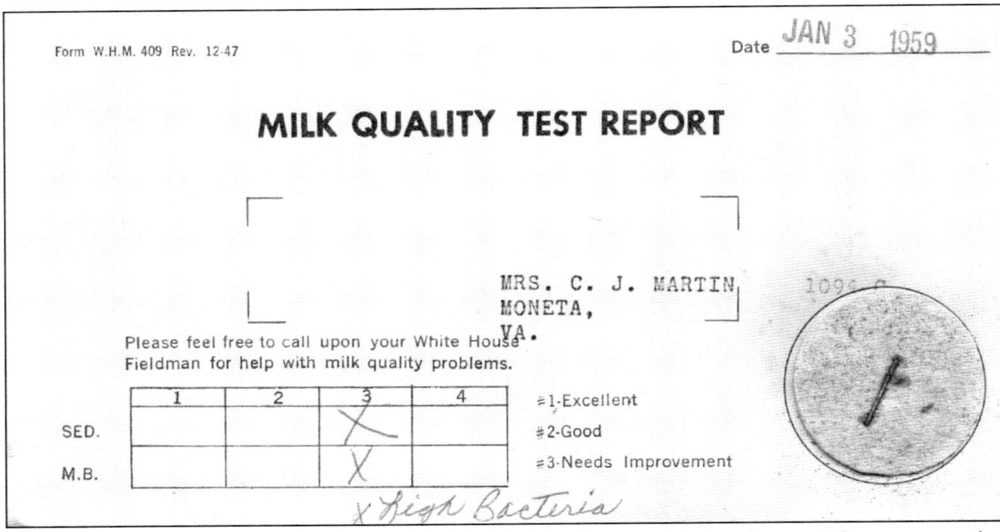
Form W.H.M. 409 Rev. 12-47

Date JAN 3 1959

MILK QUALITY TEST REPORT

MRS. C. J. MARTIN,
MONETA,
VA.

Please feel free to call upon your White House Fieldman for help with milk quality problems.

	1	2	3	4
SED.			X	
M.B.			X	

#1-Excellent
#2-Good
#3-Needs Improvement

X High Bacteria

Each week, Coble Dairy provided a Milk Quality Test Report to producers. This report, which is dated April 10, 1959, and addressed to Harold Martin, shows a milk rating for this week as a 3, and "needs improvement." Martin saved almost all of the milk report cards that he received during his career as a dairy farmer. Most, however, were better than this one. (Courtesy of the author.)

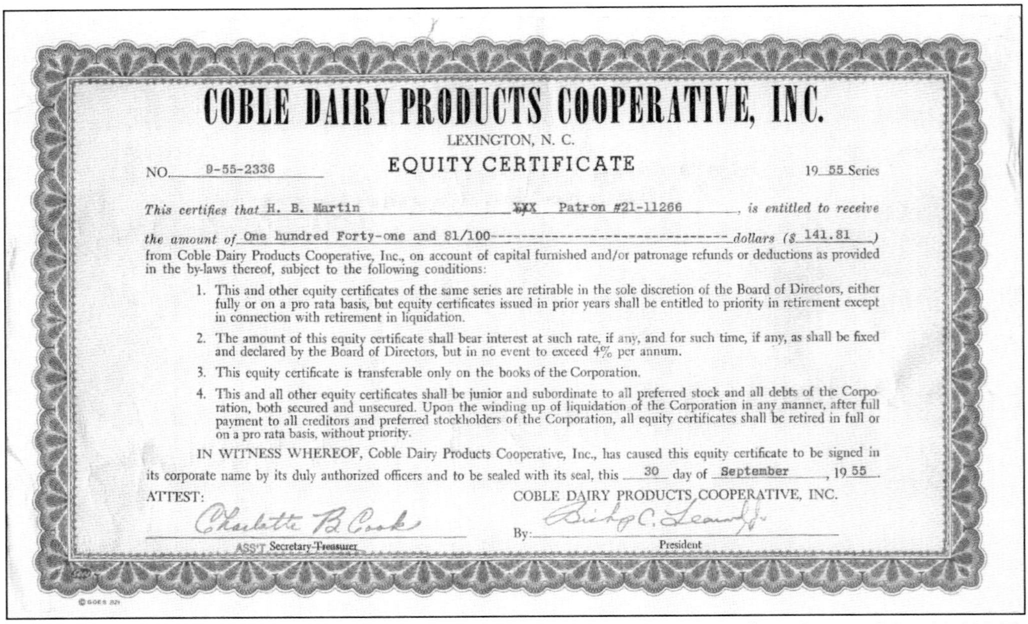
COBLE DAIRY PRODUCTS COOPERATIVE, INC.

LEXINGTON, N. C.

NO. 9-55-2336 EQUITY CERTIFICATE 19 55 Series

This certifies that H. B. Martin XXX Patron #21-11266, *is entitled to receive the amount of* One hundred Forty-one and 81/100 *dollars* ($ 141.81) from Coble Dairy Products Cooperative, Inc., on account of capital furnished and/or patronage refunds or deductions as provided in the by-laws thereof, subject to the following conditions:

1. This and other equity certificates of the same series are retirable in the sole discretion of the Board of Directors, either fully or on a pro rata basis, but equity certificates issued in prior years shall be entitled to priority in retirement except in connection with retirement in liquidation.
2. The amount of this equity certificate shall bear interest at such rate, if any, and for such time, if any, as shall be fixed and declared by the Board of Directors, but in no event to exceed 4% per annum.
3. This equity certificate is transferable only on the books of the Corporation.
4. This and all other equity certificates shall be junior and subordinate to all preferred stock and all debts of the Corporation, both secured and unsecured. Upon the winding up of liquidation of the Corporation in any manner, after full payment to all creditors and preferred stockholders of the Corporation, all equity certificates shall be retired in full or on a pro rata basis, without priority.

IN WITNESS WHEREOF, Coble Dairy Products Cooperative, Inc., has caused this equity certificate to be signed in its corporate name by its duly authorized officers and to be sealed with its seal, this 30 day of September, 19 55.

ATTEST: COBLE DAIRY PRODUCTS COOPERATIVE, INC.

Charlotte B. Cook By: [signature]

ASS'T Secretary-Treasurer President

This 1955 series Coble Dairy Products Equity Certificate was issued to Patron No. 21-11266, Harold Martin, for $141.81. The purpose was "on account of capital furnished and/or patronage refunds or deductions as provided in the by-laws." (Courtesy of the author.)

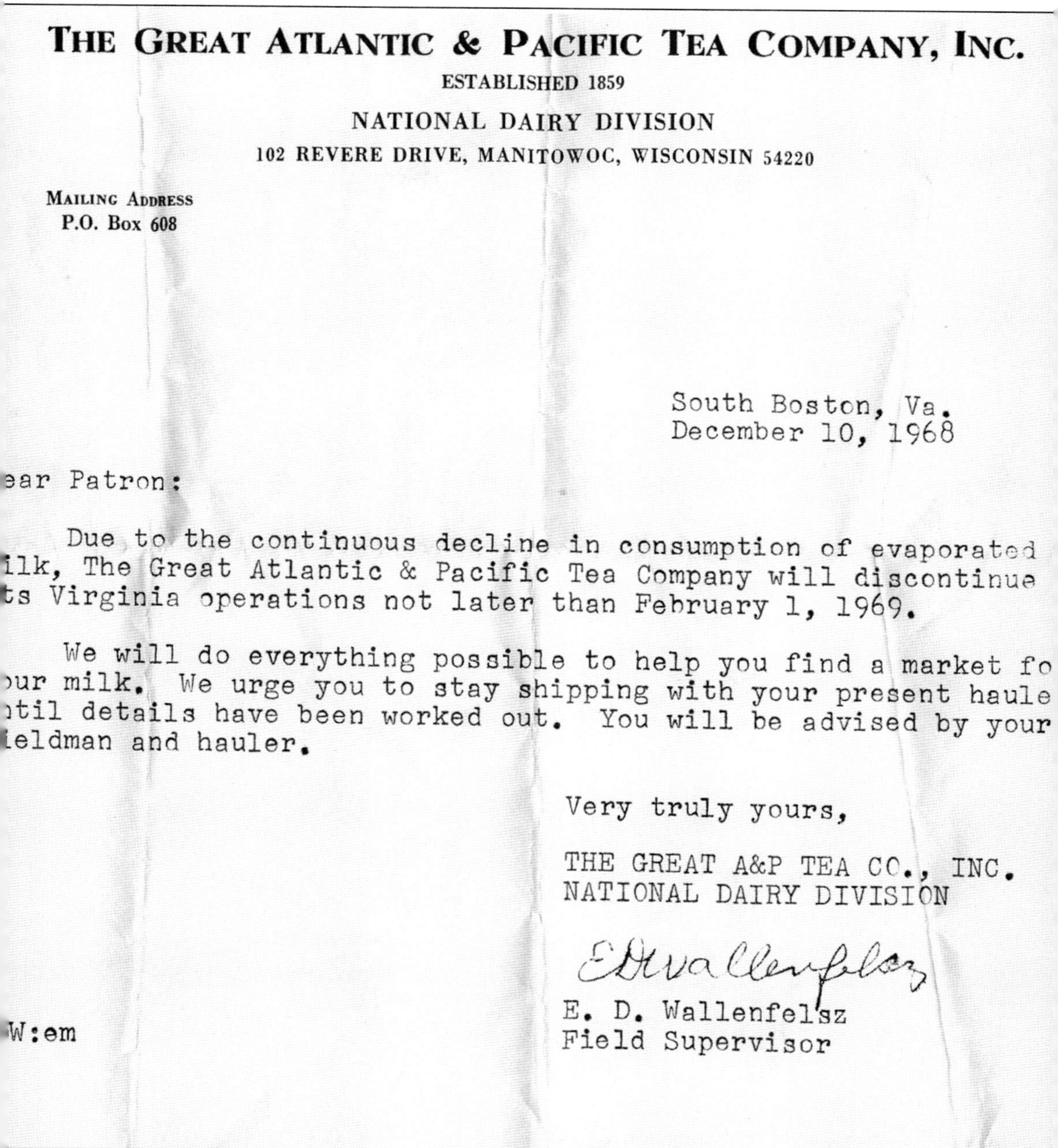

THE GREAT ATLANTIC & PACIFIC TEA COMPANY, INC.

ESTABLISHED 1859

NATIONAL DAIRY DIVISION

102 REVERE DRIVE, MANITOWOC, WISCONSIN 54220

MAILING ADDRESS
P.O. Box 608

South Boston, Va.
December 10, 1968

ear Patron:

Due to the continuous decline in consumption of evaporated ilk, The Great Atlantic & Pacific Tea Company will discontinue ts Virginia operations not later than February 1, 1969.

We will do everything possible to help you find a market fo our milk. We urge you to stay shipping with your present haule ntil details have been worked out. You will be advised by your ieldman and hauler.

Very truly yours,

THE GREAT A&P TEA CO., INC.
NATIONAL DAIRY DIVISION

E. D. Wallenfelsz
Field Supervisor

W:em

This December 10, 1968, letter from the Great Atlantic and Pacific Tea Company, Inc., is a harbinger of things to come; soon the Coble Transfer Station in Moneta would be closed. This letter informs the producer that a "continuous decline in consumption of evaporated milk will lead the company to close their Virginia operations." (Courtesy of the author.)

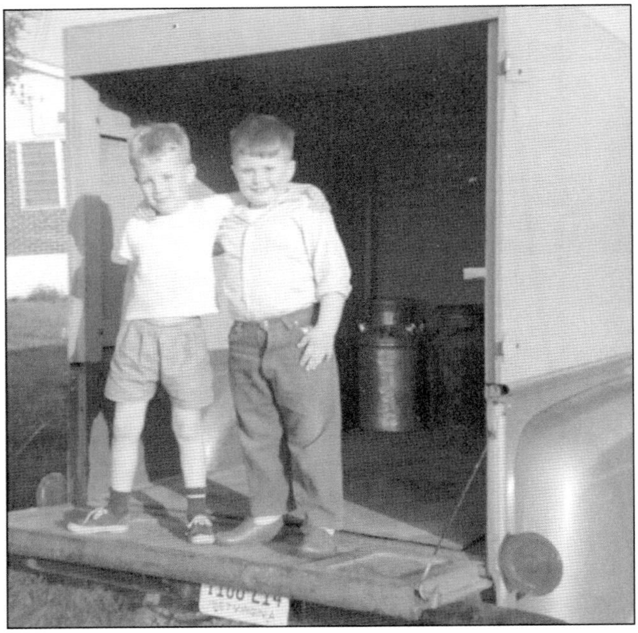

Ben Martin (right) and his cousin, Warren Saunders, ages six and five respectively, are standing in the back of an International truck Harold Martin used to haul milk cans to the Coble Transfer Station in Moneta. Several empty milk cans are visible in the rear of the truck. (Courtesy of the author.)

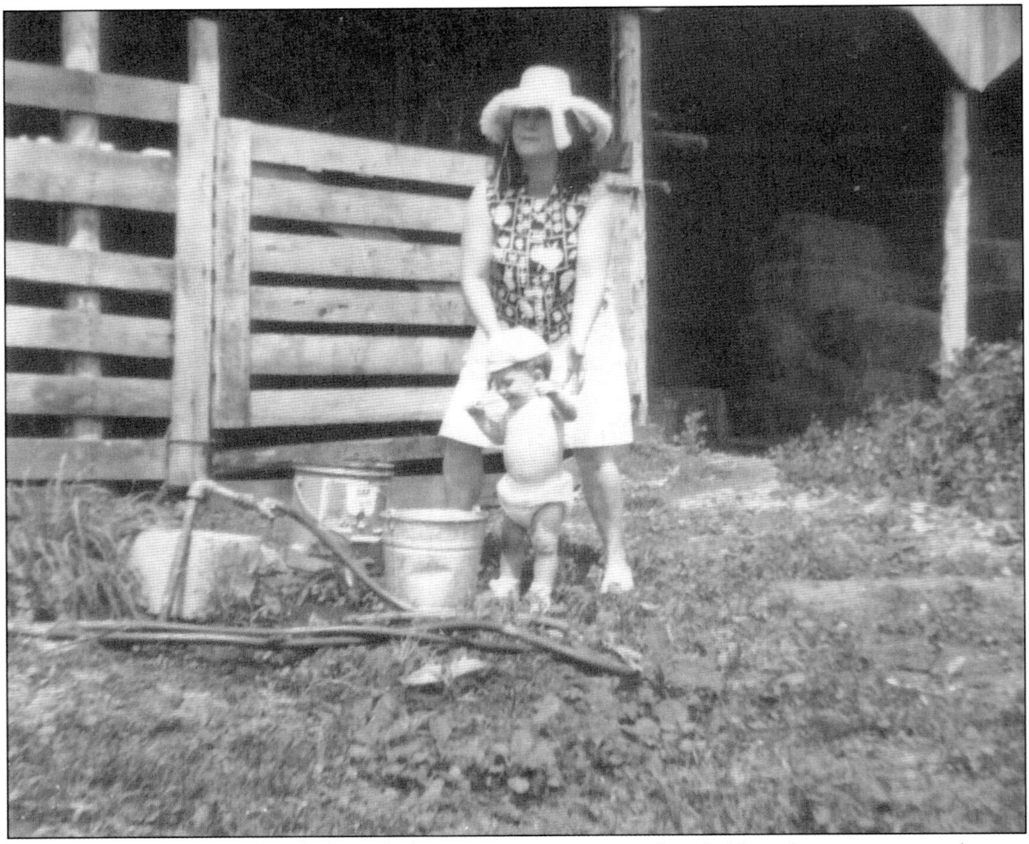

Apparently Gladys Arthur believed that proper instruction for children began at an early age. Here in 1965, she seems to be teaching her grandson how to water the livestock on the Kasey Seats farm. (Courtesy of the author.)

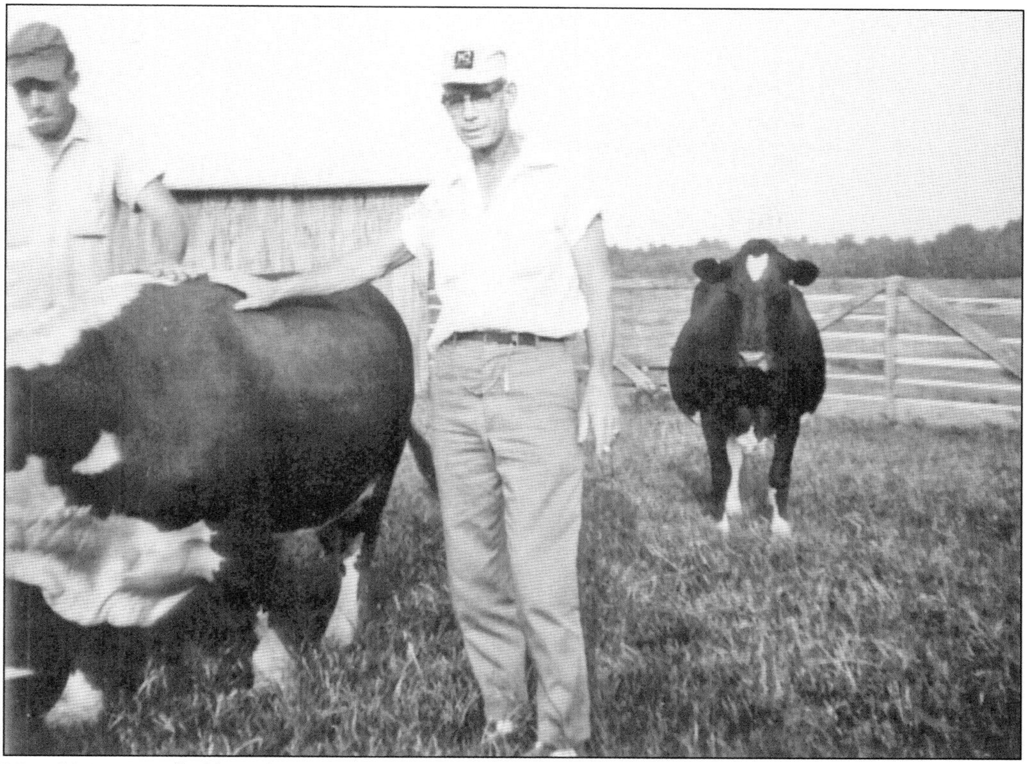

Harold Martin (left) and A. W. "Doc" Arthur are inspecting the livestock on the Kasey Seats Farm in this 1965 photograph. (Courtesy of the author.)

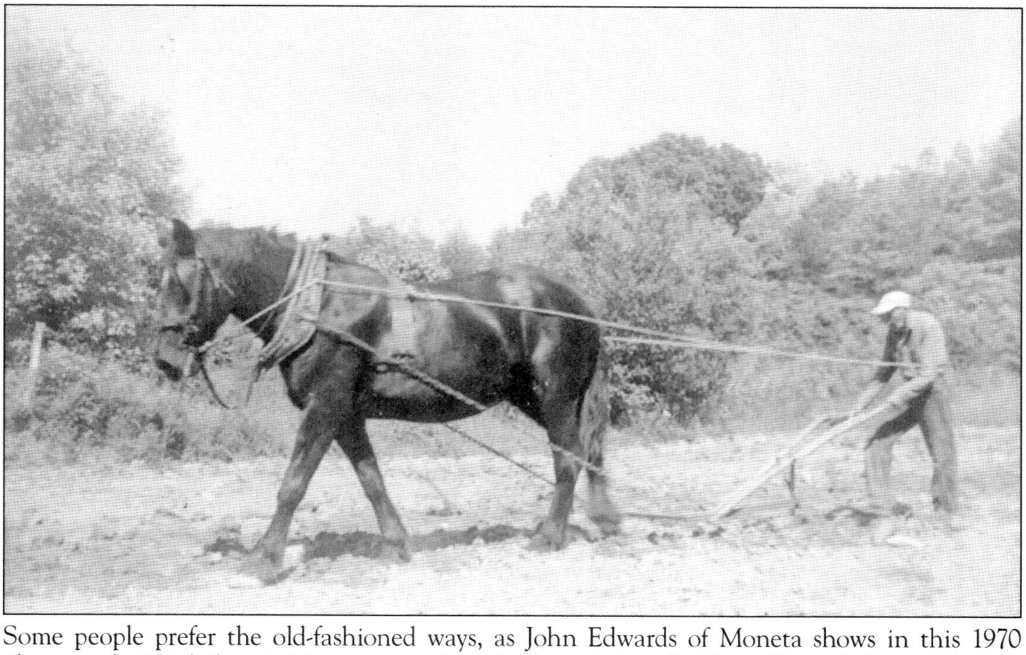

Some people prefer the old-fashioned ways, as John Edwards of Moneta shows in this 1970 photograph. His helper is Nancy, owned by the late Aubrey Bobbitt. She is helping Edwards plow his garden near the present-day Smith Mountain Lake Animal Hospital. (Courtesy of Shirley Ring.)

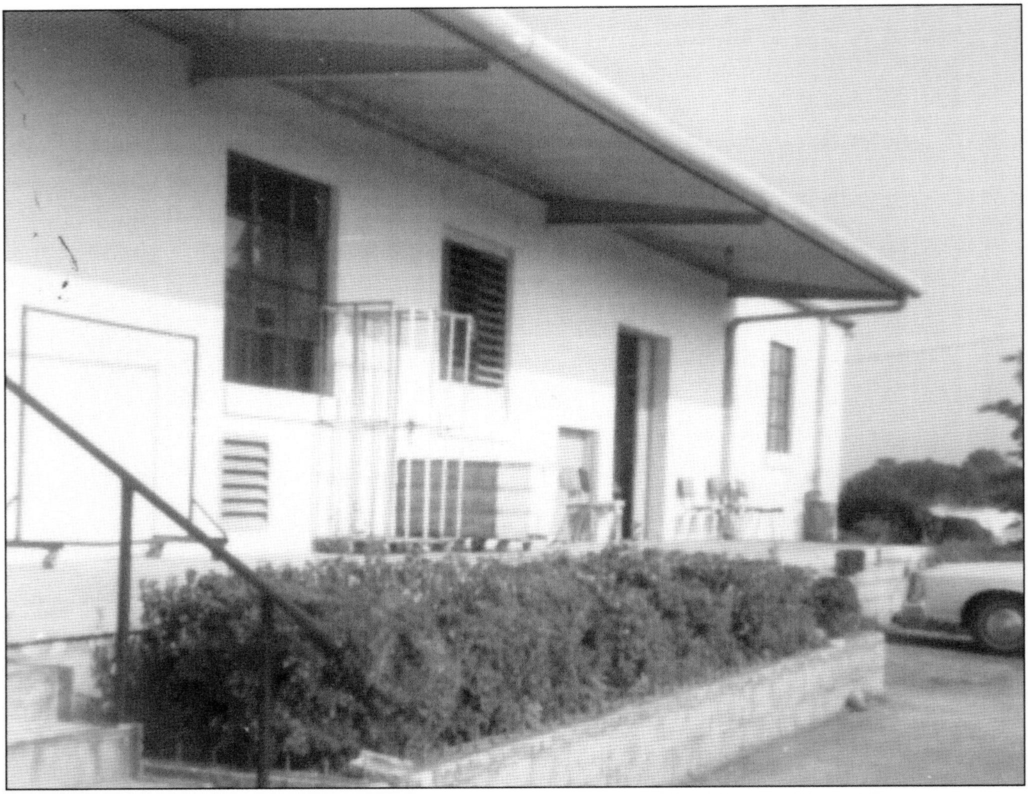

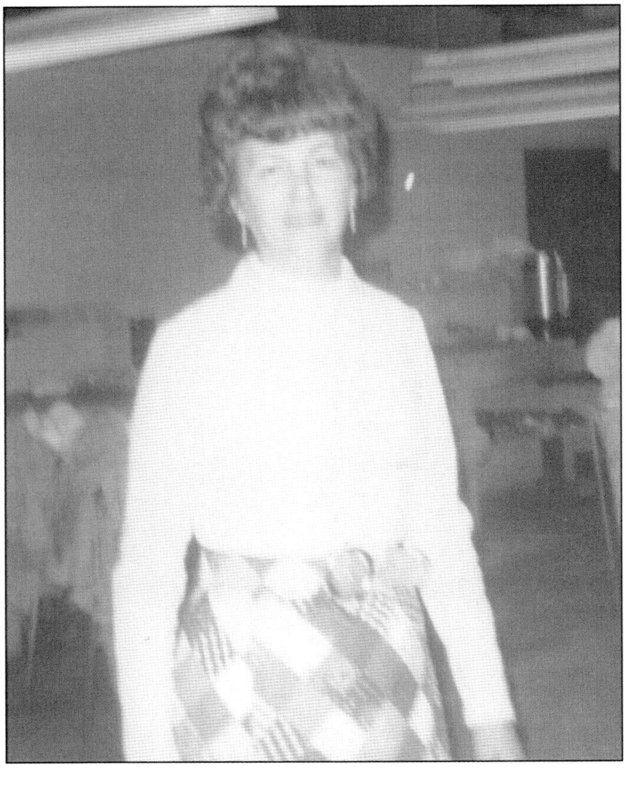

The Moneta Dress Factory building is still standing on Promised Land Road in Moneta. The Halmode Apparel Company, which is now a subsidiary of Kellwood, Inc., completed this structure on December 7, 1966. According to *Bedford Villages Lost and Found*, the company employed around 190 employees at one time, including many Bedford County farm wives. (Courtesy of Shirley Ring.)

Jean Mullens, who was the Moneta Dress Factory plant manager, poses for the camera in this undated photograph inside the plant. (Courtesy of Shirley Ring.)

Carolyn Johnson, who was a manager in the finishing department, takes a break during the work day. (Courtesy of Shirley Ring.)

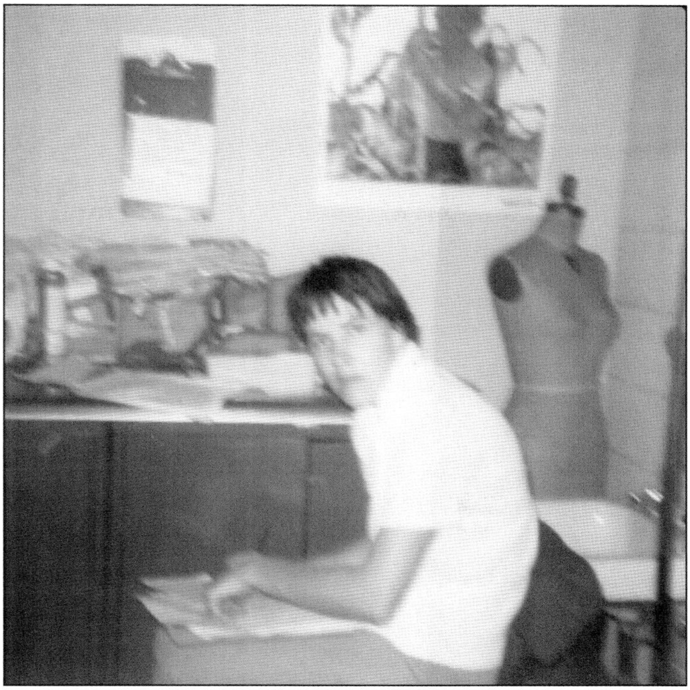

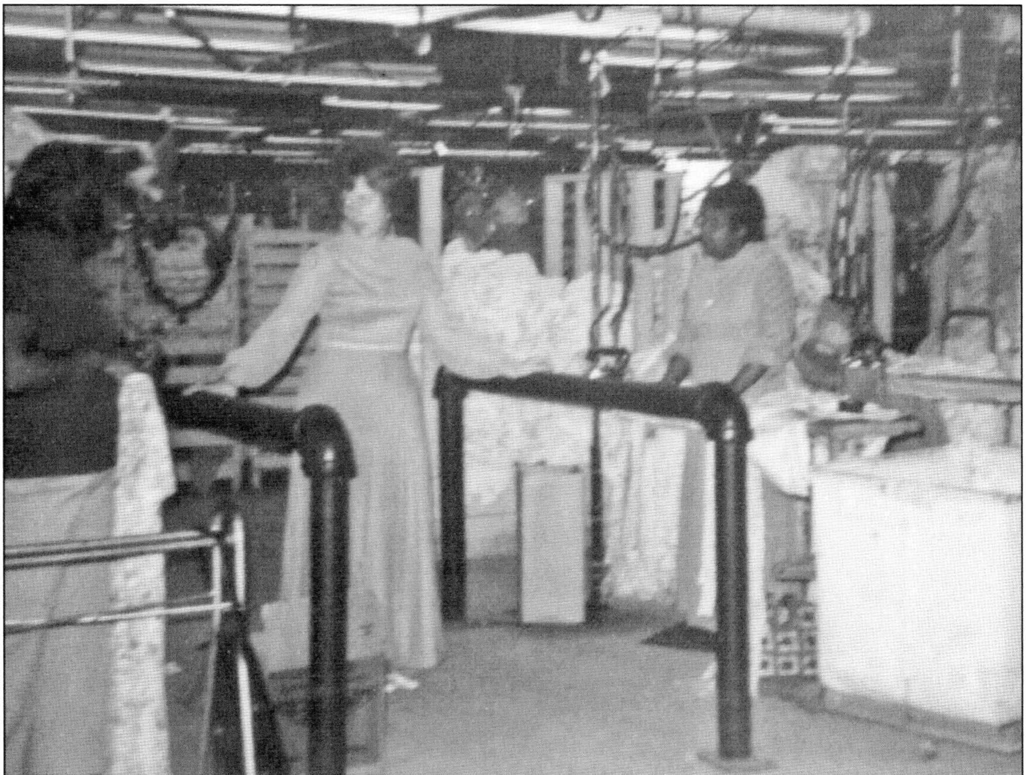

Here are busy workers inside the finishing department of the Moneta Dress Factory. (Courtesy of Shirley Ring.)

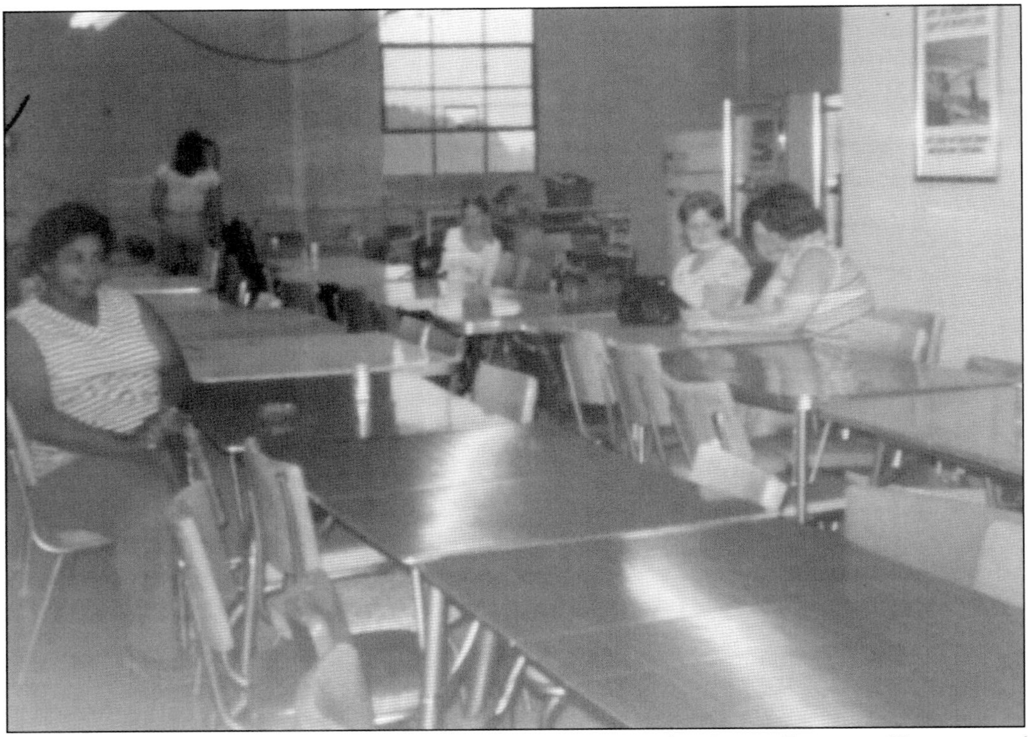

Workers are enjoying lunch inside the break room at the Dress Factory. (Courtesy of Shirley Ring.)

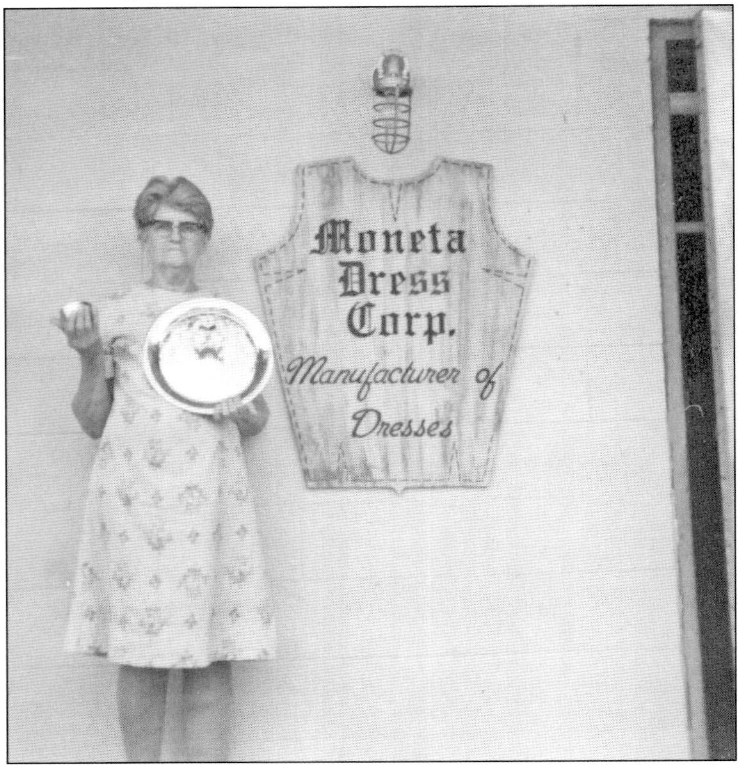

Ruby Stover Forbes Edwards receives a pewter cup and plate upon her retirement in 1979 from the Moneta Dress Factory. Edwards had 22 years of service with the company. (Courtesy of Shirley Ring.)

Three

Country Folk

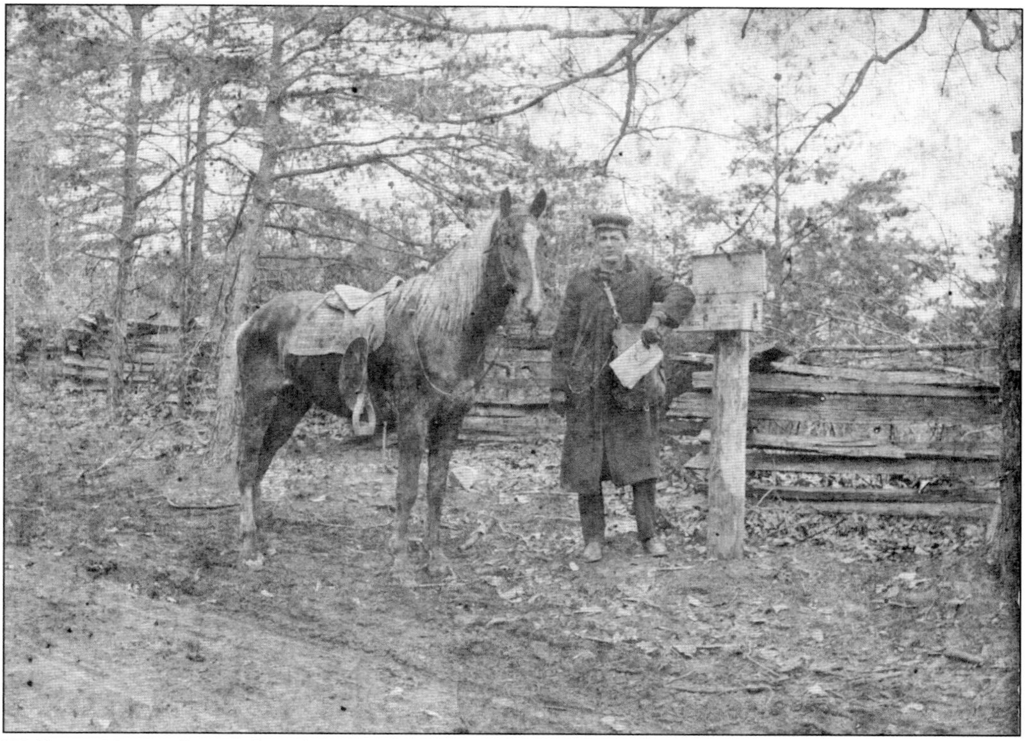

William Howard Meador delivers mail in 1920s Moneta in this photograph. This was his occupation prior to building a general store in Moneta that would, decades later, be used in a Hollywood film. (Courtesy of Gladys Hoover.)

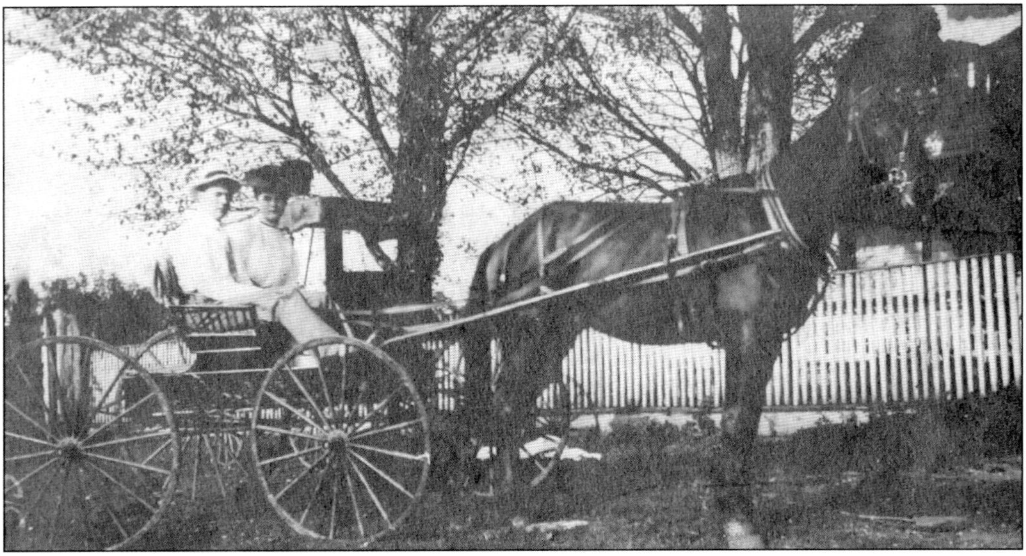

A young Howard Meador and Lena Bradley are preparing for an afternoon of courting in this photograph from the early 1900s. (Courtesy of Gladys Hoover.)

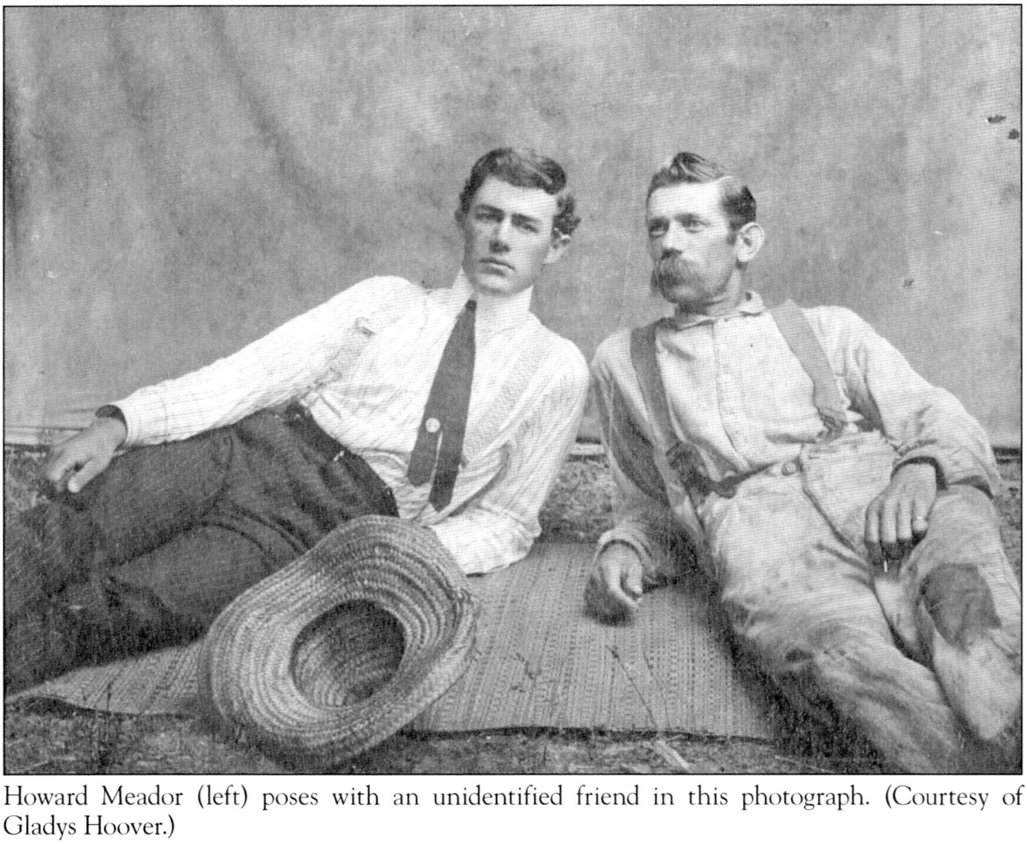

Howard Meador (left) poses with an unidentified friend in this photograph. (Courtesy of Gladys Hoover.)

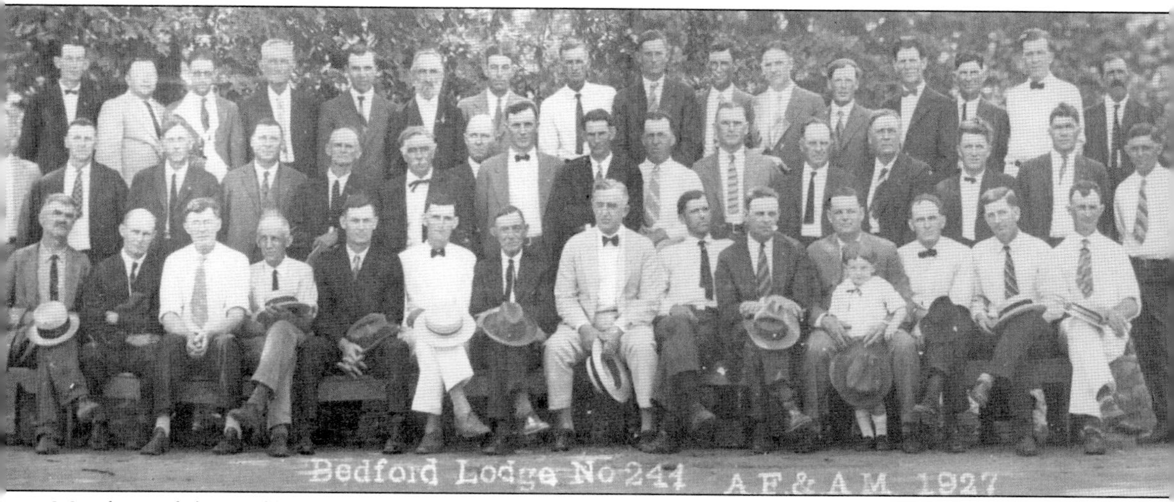

Members of the Bedford Lodge No. 244 pose in 1927. The members are, from left to right, (first row) E. M. ("Elmer") Quisenberry, Warren L. Updike, William Howard Meador, John Forbes, Rad ("G. R.") Nance, Kane Watson, Jim Dinwiddie, Judge Clifton, Henry Bond, Moorman Johnson, Frank Johnson, Frank Johnson Jr., Mack H. Updike, H. E. Nance, and William B. Cheek Sr.; (second row) Jim F. Nance, Luther Scott, Nathan M. Blackwell, Hampton Johnson, Joe Tuck, Jim Leftwich, Gaston Updike, E. L. ("Dick") Peters, Gilbert Hicks, Jack Hubbard, Donnie Lipscomb, Walter Johnson, Oliver Bays, Robert Holdren, Frank Hubbard, and Pete Holland Jr.; (third row) Owen Chambers, Jimmy Thompson, Robert Johnson, C. C. ("Kit") Cheek, Harry Overstreet, D. W. ("Dave") Parker, Homer Burnette, Mack Johnson, Rucker Witcher, Gil Anderson, S. W. ("Pete") Woodford, Otie Mitchell, Earnest Mitchell, Sam Anderson, Wallace Dinwiddie, and Orie Patterson. (Courtesy of Gladys Hoover.)

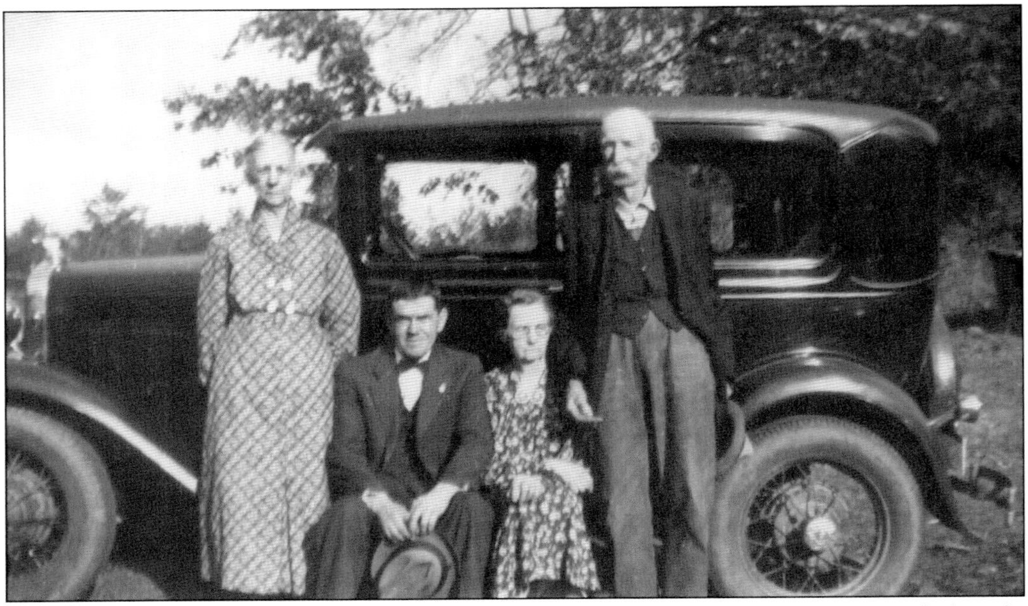

Mr. and Mrs. Covey Forbes (standing) pose beside their family car during a trip with friends. (Courtesy of Shirley Ring.)

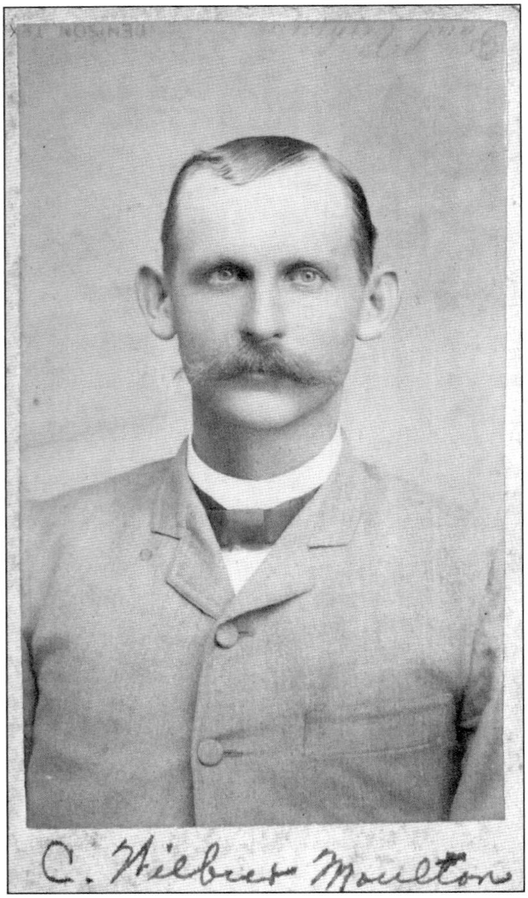

C. Wilbur Moulton, a relative of Dr. Benjamin H. Moulton, who was a Bedford County Civil War–era doctor, is pictured here. Dr. Moulton, according to *Bedford Villages Lost and Found* (compiled by the Peaks of Otter Chapter of the DAR) came from Maine to teach at a school in the Bunker Hill area of Bedford County. The log cabin from which first he, then later, Dr. Sam Rucker Sr., practiced, still stands on State Route 122 near Goose Creek Bridge. Legend has it that during the Civil War, when the Northern army retreated through Bedford County and discovered that Moulton was a doctor, they not only forced him to treat their wounded, but they stole his horse when they left. (Courtesy of Dr. James Rucker Jr.)

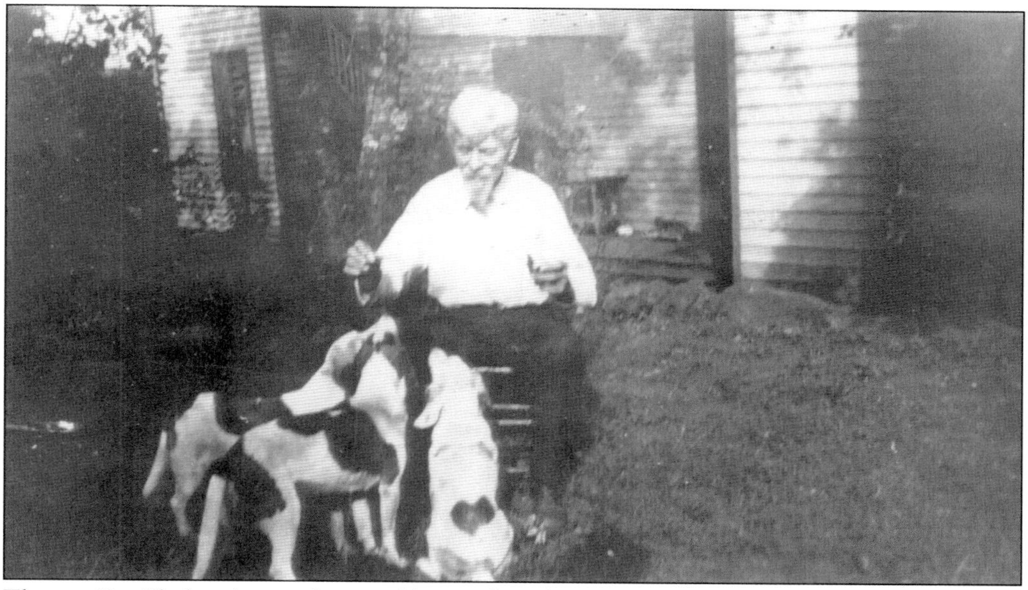

Thomas Key Phelps plays with several loving friends in this photograph taken shortly before his death in 1925. A Virginia Military Institute keydet, he was a veteran of the Civil War and fought at the Battle of New Market. (Courtesy of Dr. James Rucker Jr.)

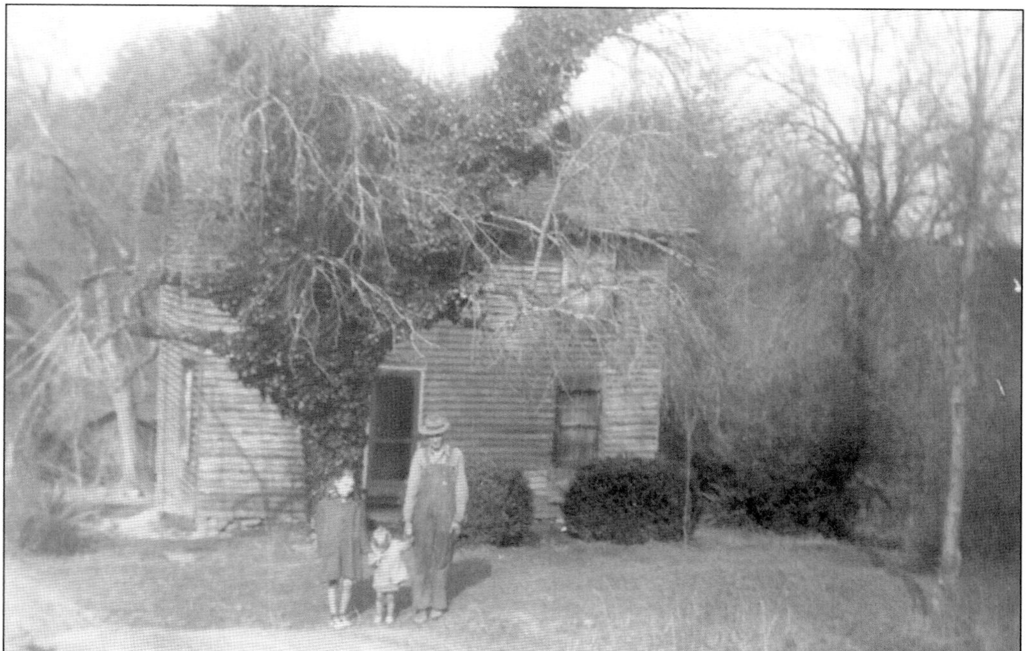

This home, which belonged to John Edwards, was located on the road that would become State Route 122 near its intersection with Hendricks Store Road. Posing here in the 1940s, from left to right, are Betty Jean, Shirley, and John Edwards. (Courtesy of Shirley Ring.)

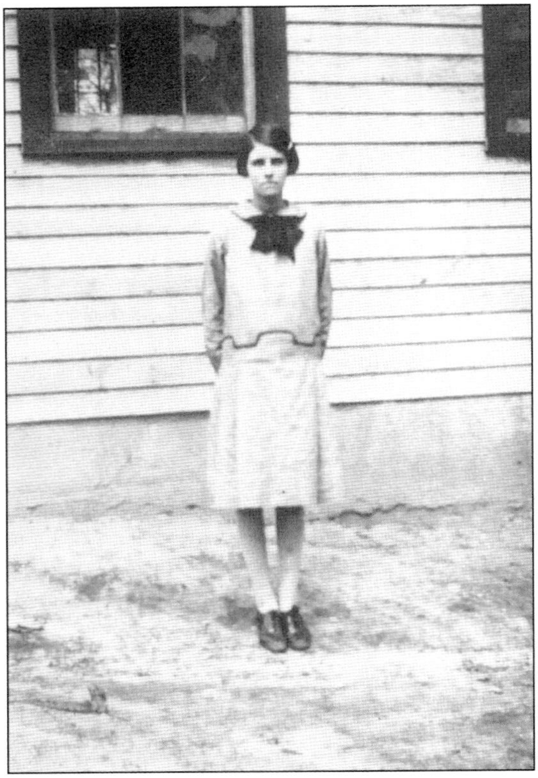

Ruby Stover Forbes Edwards poses outside the Western Light School that was located near the present-day Sunrise Builders on State Route 122. In later years, the school would be moved to Hendricks Store Road and renamed Promised Land School. Here the school would serve young African American children during the segregation era. (Courtesy of Shirley Ring.)

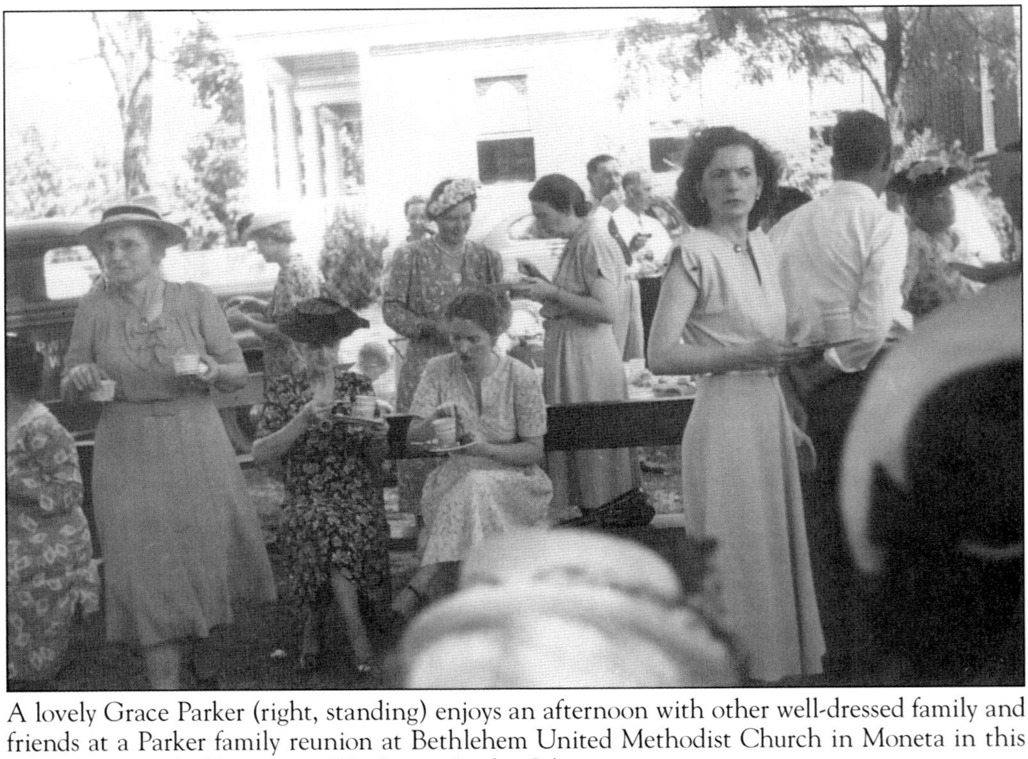

A lovely Grace Parker (right, standing) enjoys an afternoon with other well-dressed family and friends at a Parker family reunion at Bethlehem United Methodist Church in Moneta in this 1948 photograph. (Courtesy of Dr. James Rucker Jr.)

Mattie Poteat is standing outside what was the home of the Alvin Fielder family. Fielder operated a general store near the present-day Shoprite Grocery. The Fielder home place was located across the road from Shoprite on State Route 122 in Moneta. (Courtesy of Shirley Ring.)

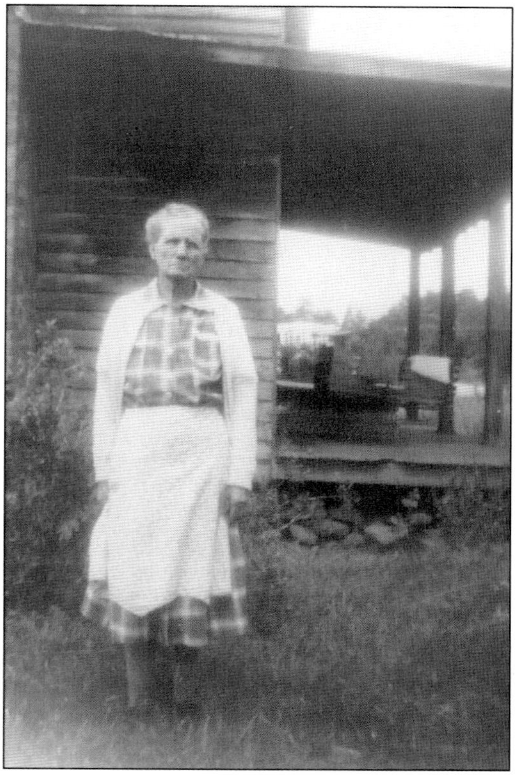

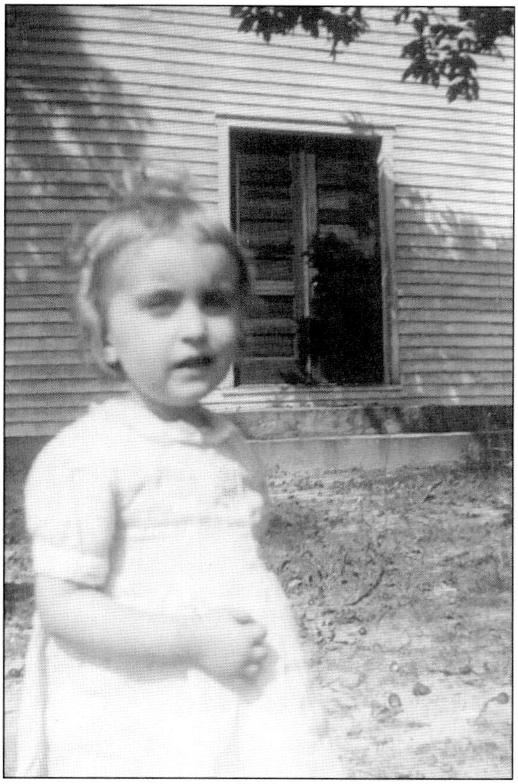

This darling toddler is Betty Jean Edwards Blankenship, standing outside the Western Lake Tabernacle Church, which was located on a much earlier, dirt-covered version of State Route 122. (Courtesy of Shirley Ring.)

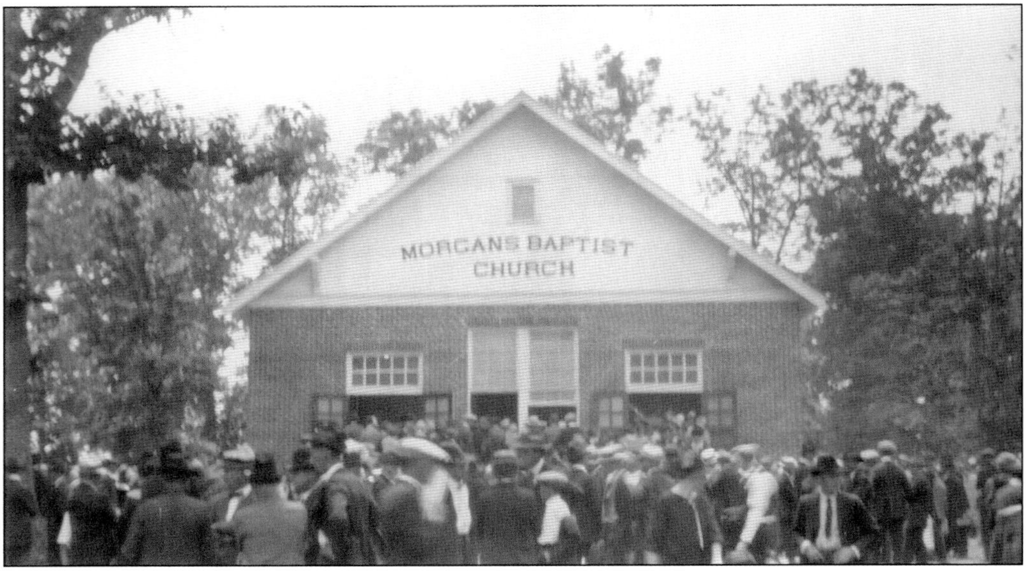

This photograph, which is on the back of a postcard and is undated, shows churchgoers at Morgans Baptist Church in Moneta. (Courtesy of Sam and Debbie Rucker.)

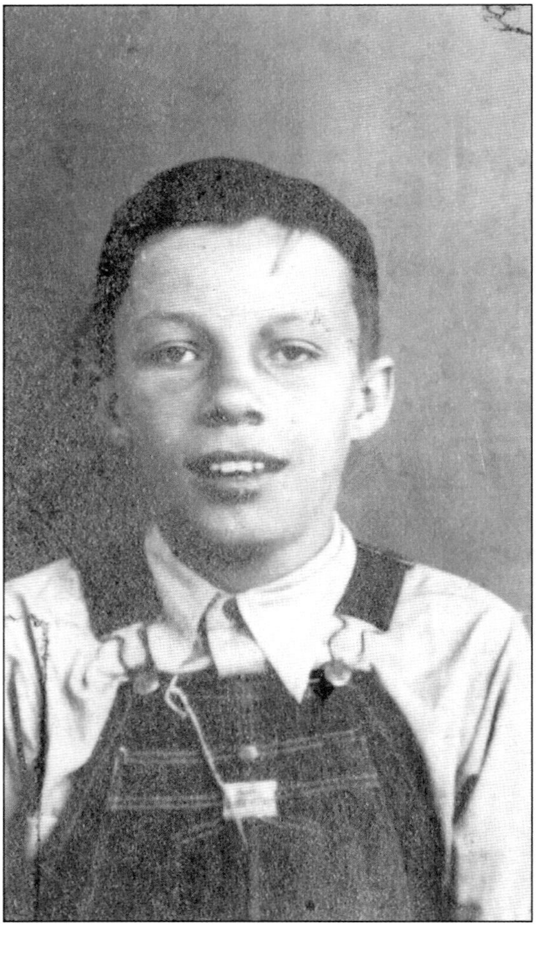

Harold Martin is almost seven years old in this photograph. Bib overalls were used for working in the fields, but a nice pair was usually saved for church services, funerals, and, as is shown here, having a picture taken. (Courtesy of the author.)

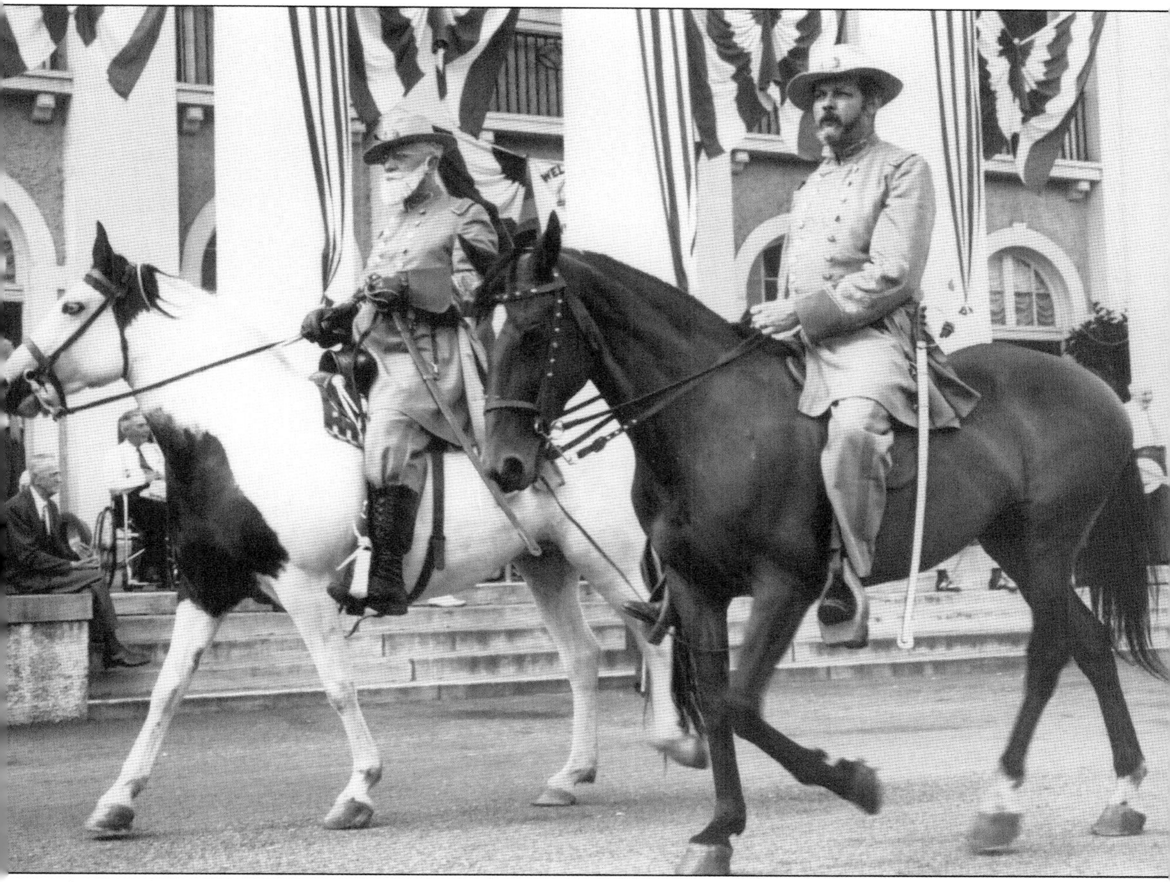

The gentleman on the white horse is William Richard Phelps, riding through downtown Bedford in 1965 during the city's centennial celebration. Phelps's white beard is also part of his uniform. (Courtesy of Sam and Debbie Rucker.)

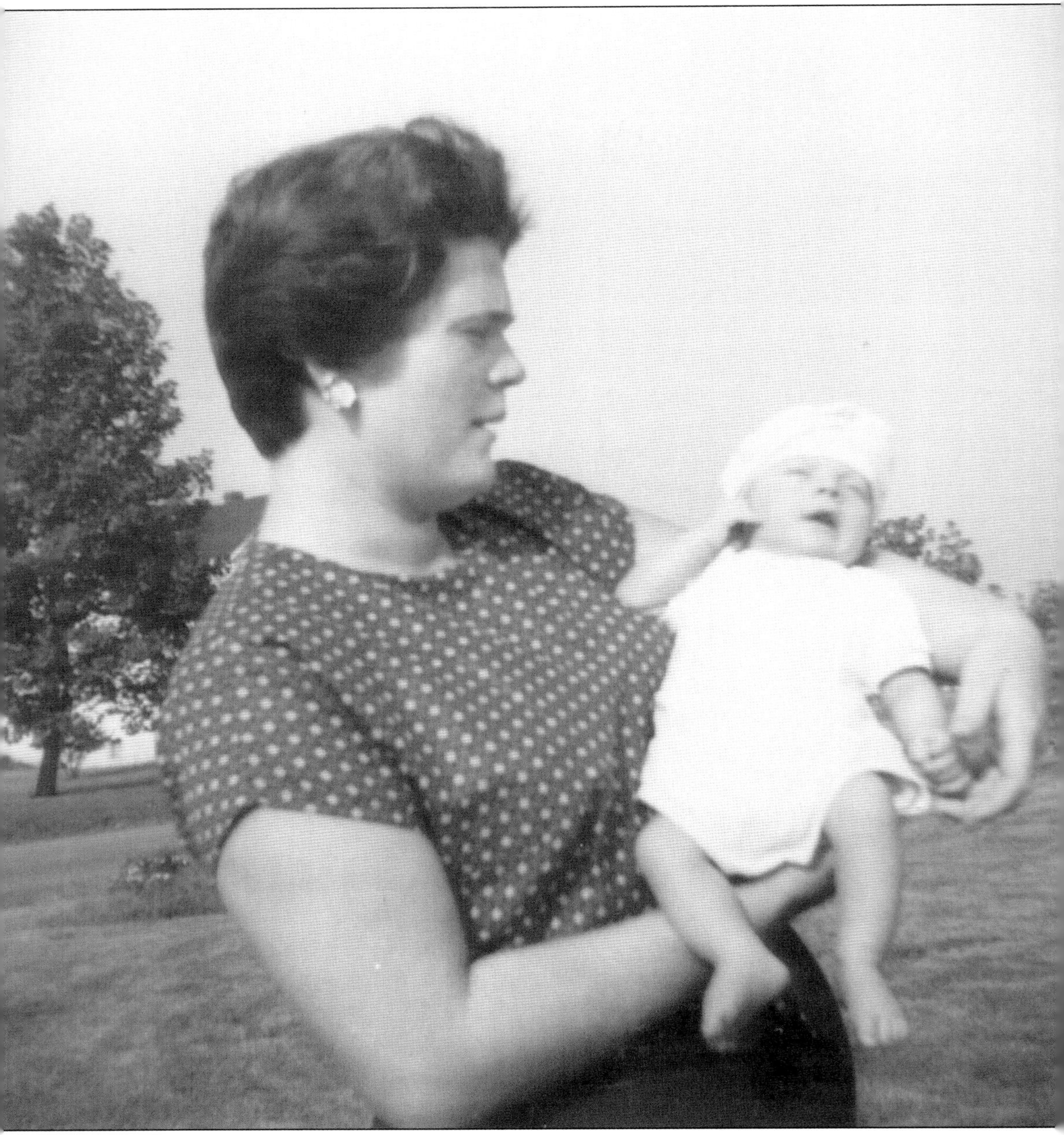

Bedford County farm wife Rose Ann Martin looks both shocked and awed by her baby son, Harold Martin Jr., in this October 1962 photograph. (Courtesy of the author.)

Four

"Dr. Sam"

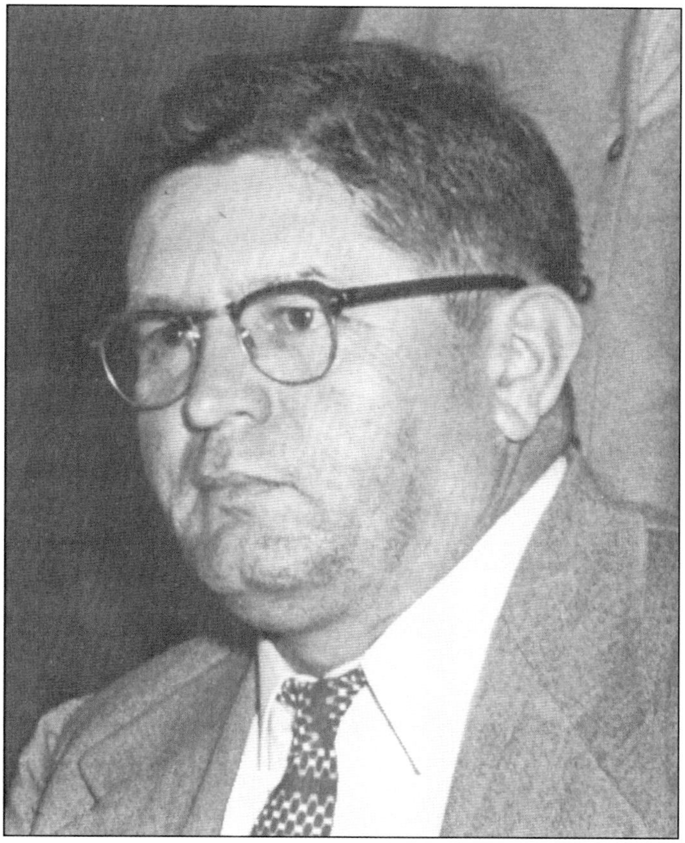

For 37 years, Dr. Sam L. Rucker Jr. practiced medicine from the large Victorian house in Moneta that was built by his father, Dr. Sam Rucker Sr., in 1912. Sam Jr. began his career in Elk View, West Virginia, and according to the August 5, 1976, *Bedford Bulletin*, had the misfortune of having the same last name as a local family of ne'er-do-wells. Once, on a house call to care for a sick child, the mother, who met the doctor at the door, was hesitant to let him enter her home for fear that her child would be afraid of a man named "Rucker." So the desperate but enterprising mother suggested he go by "Dr. Sam." This "In Appreciation" photograph comes from the 1962 Moneta High School class yearbook, the *Village*. (Courtesy of Louise Stone.)

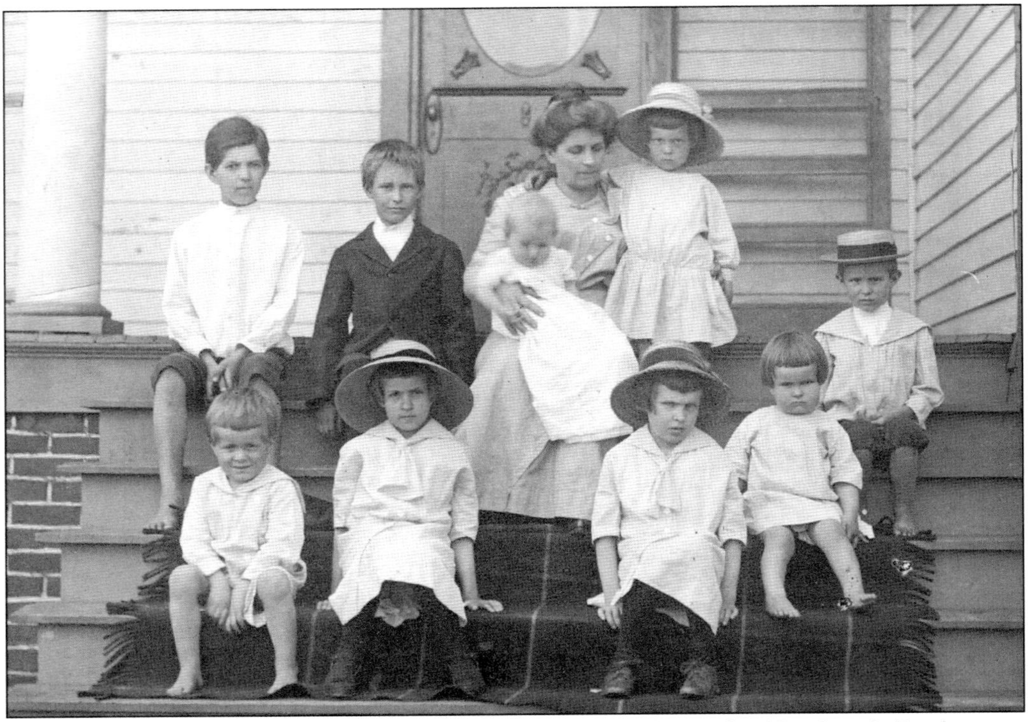

Elizabeth Rucker, wife of Dr. Sam L. Rucker Sr., a Bedford County physician for several decades prior to his son, poses on the front steps of Rucker House in Moneta with her children. From left to right are (first row) James, Elizabeth, Francis, Mary, and Sam Jr.; (second row) Edward, Frank, John (in his mother Elizabeth's arms), and Bessie. (Courtesy of Dr. James Rucker Jr.)

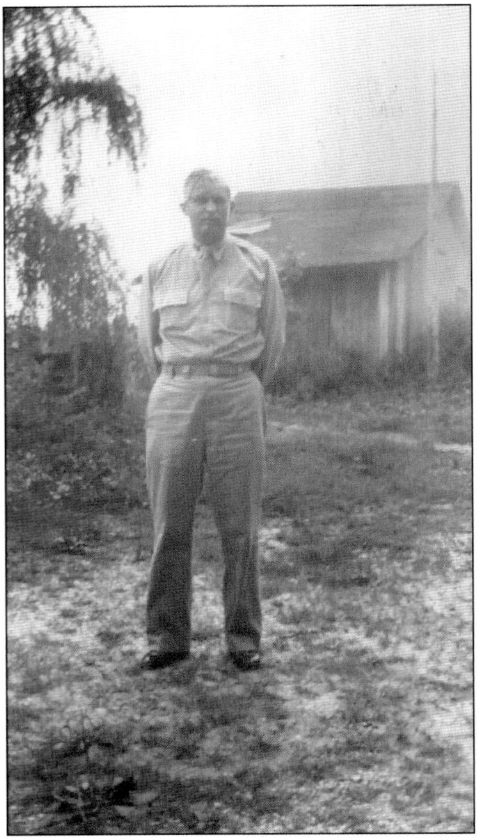

Dr. Sam Rucker Jr., who was a captain in the medical corps, is pictured here at home on a furlough. (Courtesy of Dr. James Rucker Jr.)

A young Dr. Sam Rucker Jr. (left) and one of his nephews pose with Dr. Sam Rucker Sr. are posing at Rucker House in Moneta in the late 1940s. Apparently the young man is not too unhappy to have his hand bandaged. (Courtesy of Sam and Debbie Rucker.)

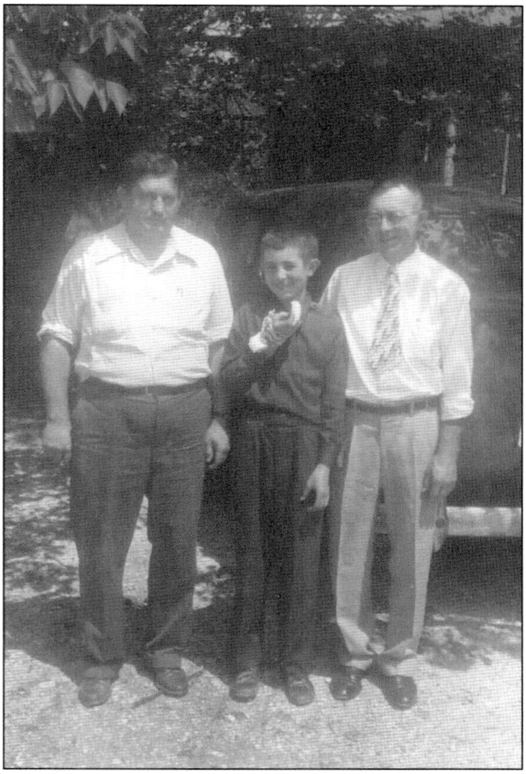

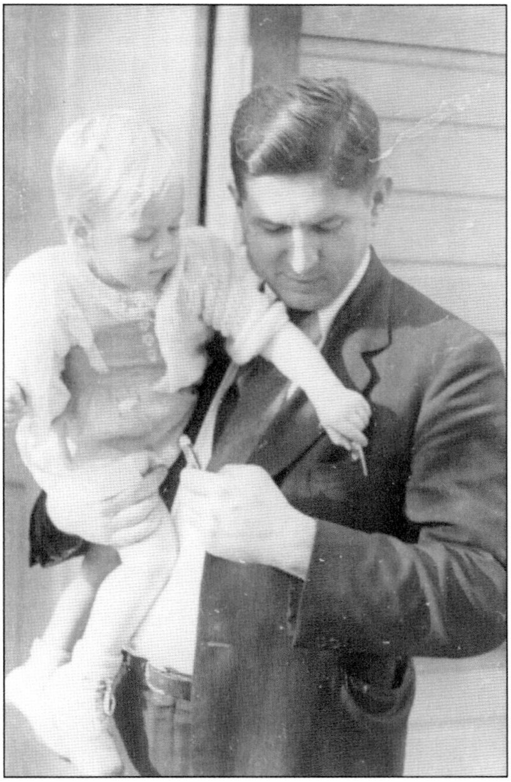

"Why in the hell didn't you call me sooner?" the cantankerous Dr. Sam bellowed into the telephone before slamming down the receiver on Rose Anne Martin, the author's mother, one Saturday night in the early 1970s; the author had developed a fever and became ill. His parents fully anticipated that the badgered, overworked, sometimes ill-tempered, but always generous country doctor would call back after he had released some steam from his pressure-cooker temper. A few minutes later, he did call back to apologize for his outburst, and she accepted the peace offering as if it had been a full-carat diamond. Dr. Rucker then proceeded to ask her a few questions before instructing her to have her husband come to his office in Moneta for some medicine. In this photograph, a young Dr. Sam is playing with his niece, Debbie Rucker. Though he could be temperamental with his adult patients, he had a great love for children. (Courtesy of Dr. James Rucker Jr.)

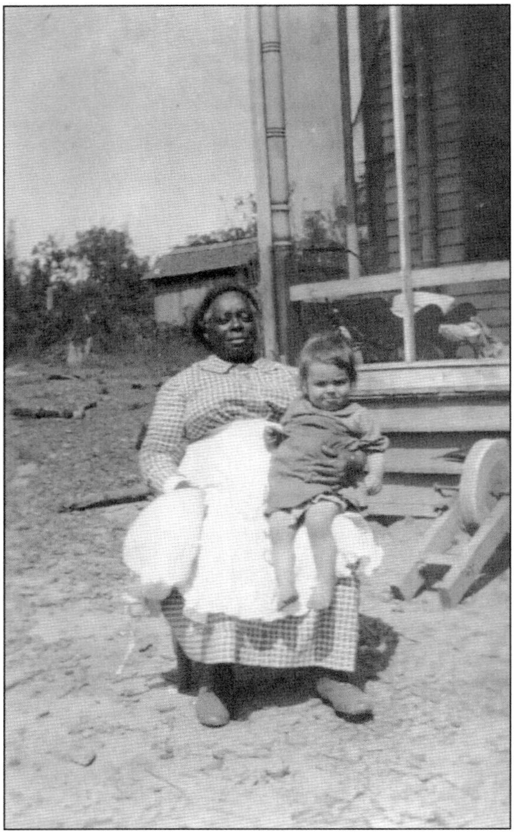

A content Richard Rucker is shown here in a 1919 photograph with his nursemaid at Rucker House. Richard would later practice dentistry in Gretna, Virginia. (Courtesy of Dr. James Rucker Jr.)

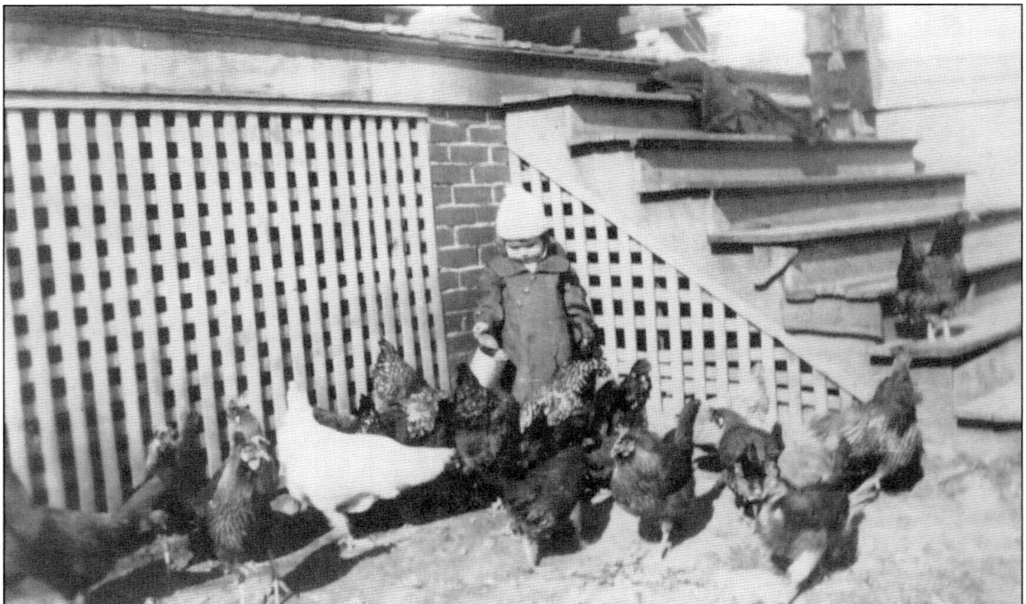

In December 1919, a puzzled Richard Rucker, shown here cornered at the front porch of Rucker House, is too young to tell his friends, "Don't peck the hand that feeds you!" (Courtesy of Sam and Debbie Rucker.)

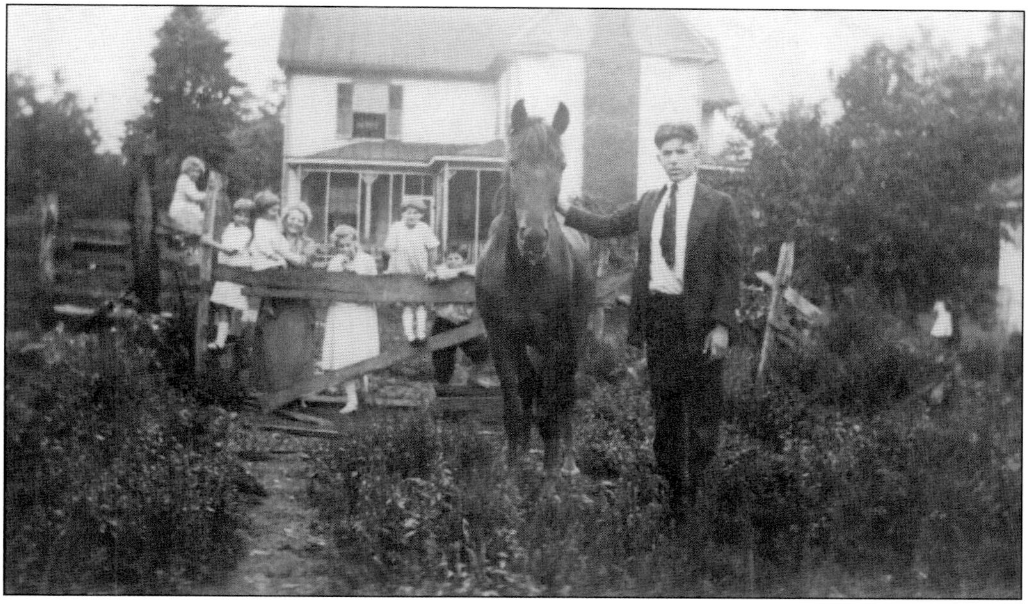

Edward Rucker and his friend Bonnie the horse pose for a September 1920 photograph. Rucker siblings and their home place are in the background. In later years, this side porch was enclosed and served as a waiting room for Dr. Sam's patients. (Courtesy of Dr. James Rucker Jr.)

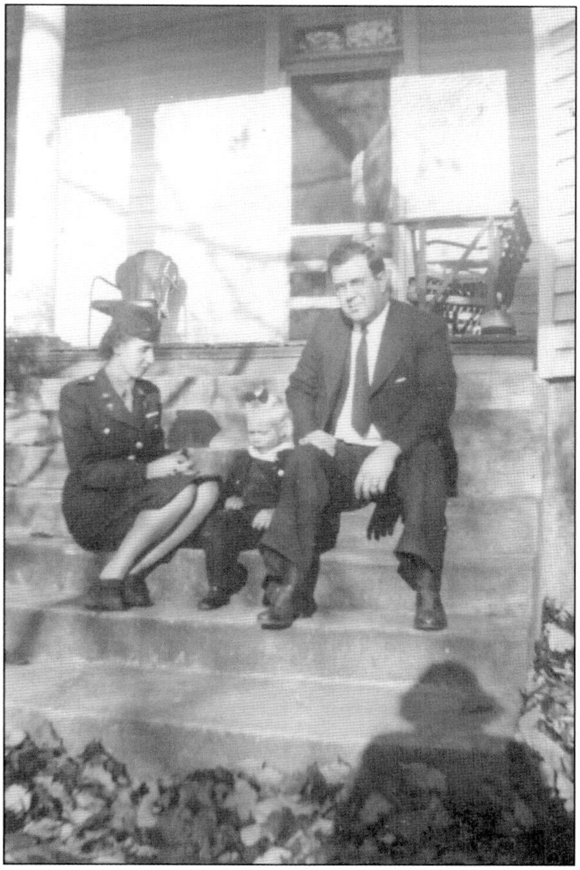

Dr. Edward Rucker, one of Dr. Sam's brothers, is on the front porch of Rucker House in Moneta with his sister, Margaret, who was in the armed forces during World War II when this photograph was taken. Edward practiced medicine in West Virginia. (Courtesy of Dr. James Rucker Jr.)

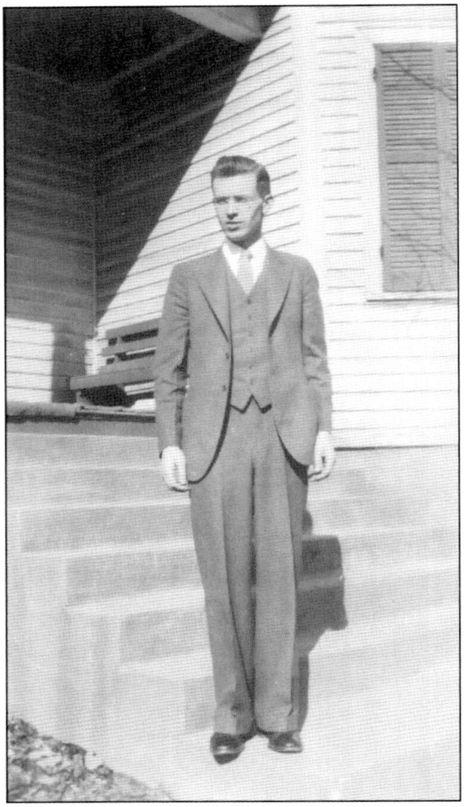

Dr. James Rucker Sr., who practiced dentistry in Rocky Mount, Virginia, in Franklin County, poses on the front steps of Rucker House in this undated photograph. His son, James Jr., still practices in the same facility he once shared with his father. (Courtesy of Dr. James Rucker Jr.)

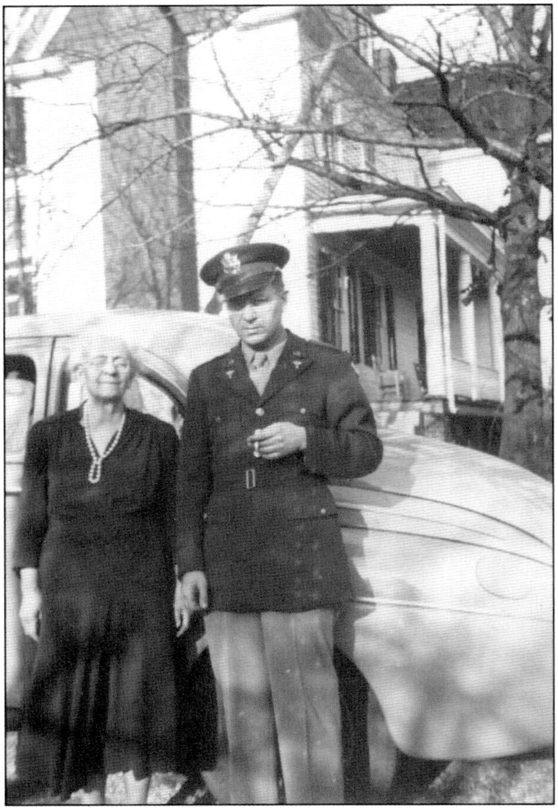

Dr. James Rucker Sr. poses with his mother, Elizabeth, during a furlough in this World War II–era photograph at Rucker House. (Courtesy of Dr. James Rucker Jr.)

Francis Rucker (left), Dr. Sam's sister, served in World War II in the Philippines and chronicled her experiences in scores of pictures that are carefully preserved in photograph albums. In later years, she served as Dr. Sam's nurse at Rucker House. She is shown with Mildred Clements and an unidentified little girl with her doll outside the Parker House on State Route 122 in Moneta in this 1921 photograph. The Parker House, which is located near the Goose Creek Bridge, is still a residence. (Courtesy of Dr. James Rucker Jr.)

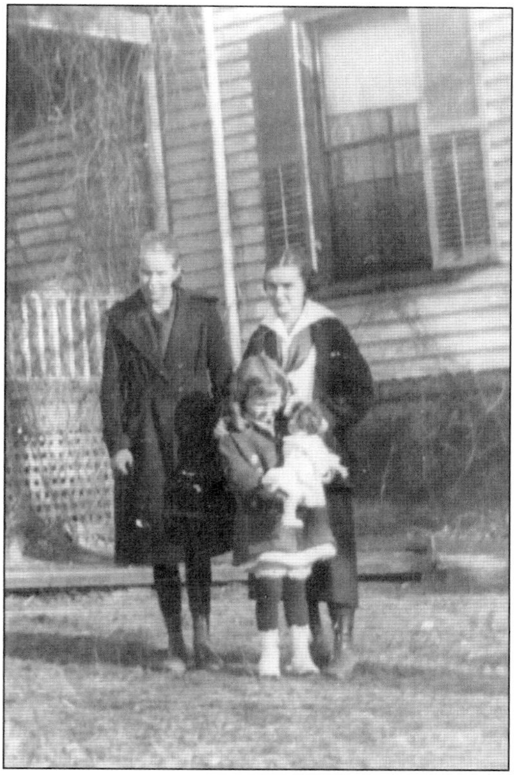

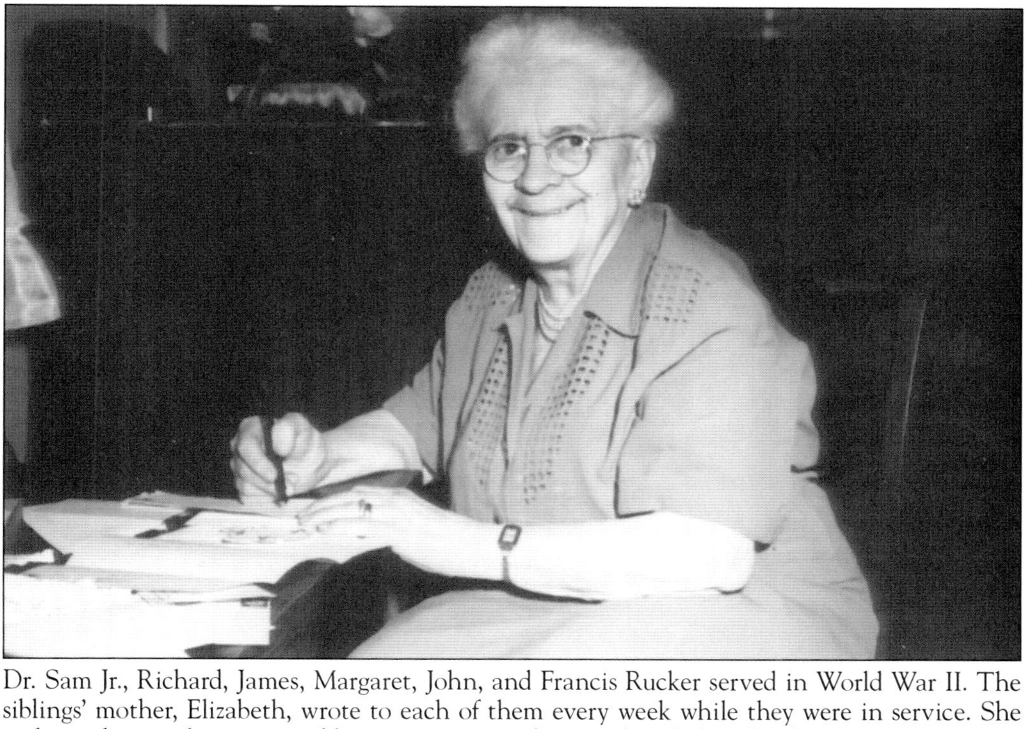

Dr. Sam Jr., Richard, James, Margaret, John, and Francis Rucker served in World War II. The siblings' mother, Elizabeth, wrote to each of them every week while they were in service. She is shown here in later years addressing correspondence at her desk in her home, Rucker House. (Courtesy of Dr. James Rucker Jr.)

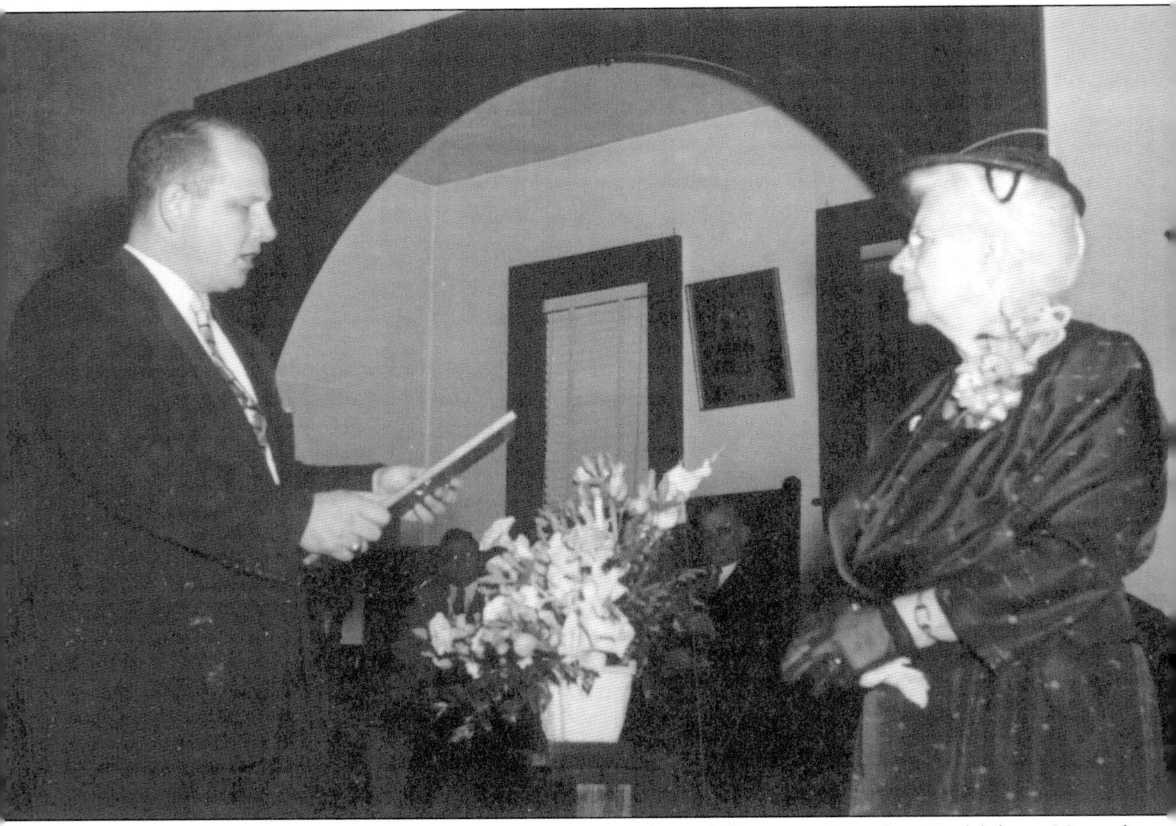

Elizabeth Margaret Rucker receives the 1956 Mother of the Year Award at Bethlehem United Methodist Church from U.S. Sixth District congressman Richard Poff. According to the online Biographical Directory of U.S. Congress, Poff was "elected as a Republican to the Eighty-third Congress; reelected to the nine succeeding Congresses and served from January 3, 1953, until his resignation August 29, 1972, to become a justice of the Supreme Court of Virginia." On an additional note, Pres. Richard M. Nixon nominated him to be justice of the Supreme Court of the United States. (Courtesy of Sam and Debbie Rucker.)

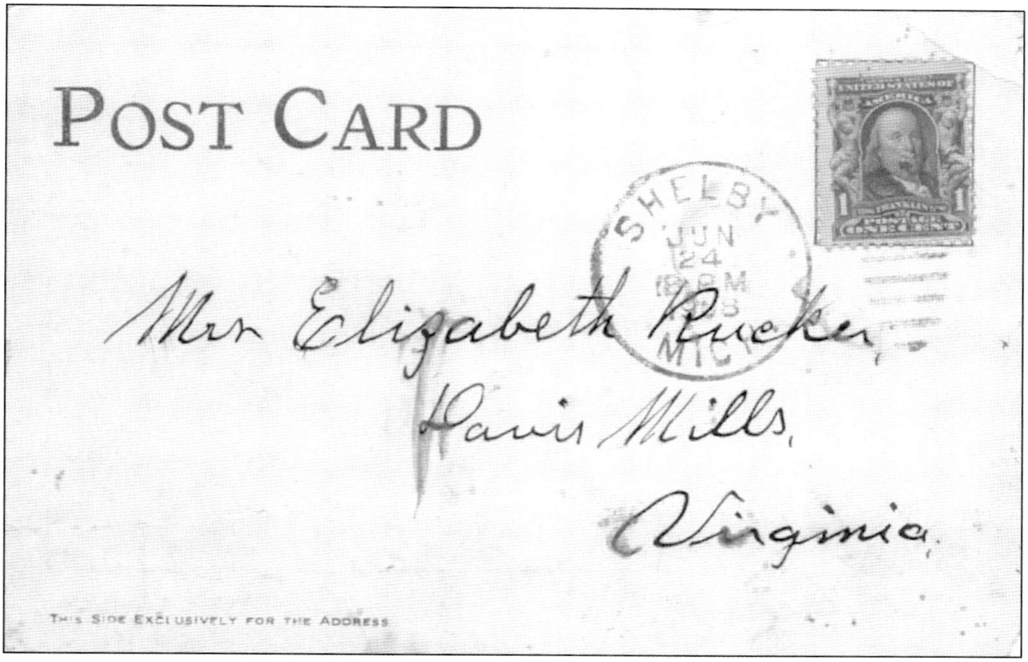

In 1908, sending a postcard to Moneta or to nearby Davis Mills was relatively easy; there were neither many Elizabeth Ruckers in the region nor many postal customers. Today uncertain writers of correspondence can "Google" an address if they are uncertain of the recipient's address. Certainly this Shelby, North Carolina, note-sender never envisioned that possibility. (Courtesy of Dr. James Rucker Jr.)

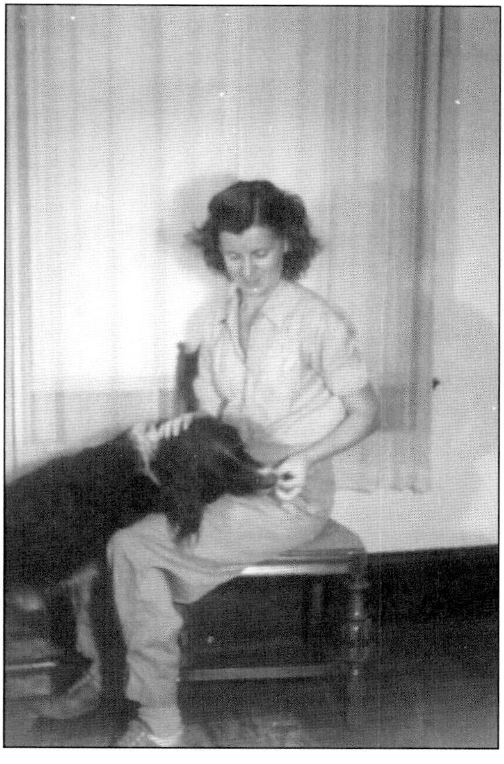

Ruth Rucker, Dr. Sam's sister, is playing with a special friend in this 1949 photograph at Rucker House. Ruth died at an early age, several years after this picture was taken. (Courtesy of Dr. James Rucker Jr.)

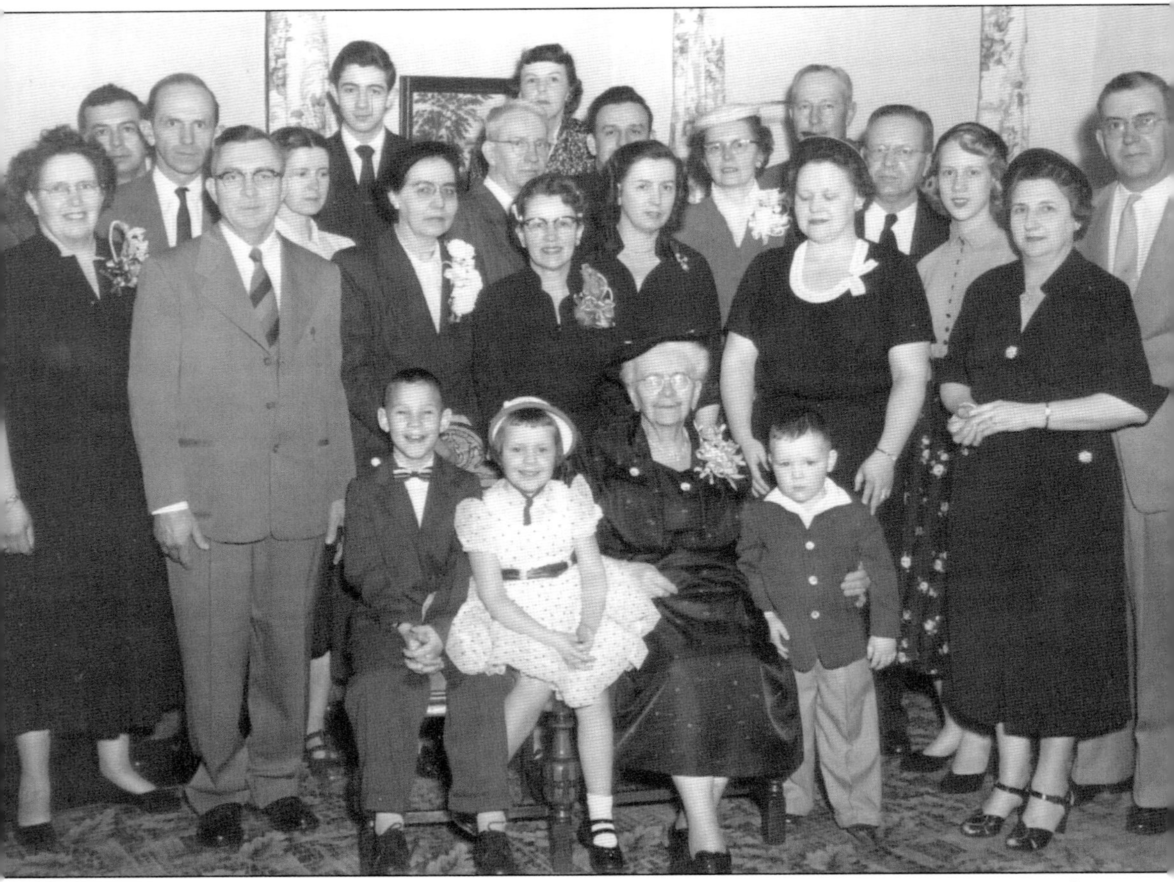

Dr. Sam Rucker Sr. (standing second row, left) and Elizabeth Rucker (first row, seated) are celebrating the marriage of their daughter, Virginia, to Carlyle Crigler in 1955 at Rucker House in Moneta. The bride is in the back row and is wearing a white hat and corsage. Her husband is behind her and to her left. The only member of the wedding party not present is Dr. Sam Jr., who was making a house call on a patient when this picture was taken. (Courtesy of Dr. James Rucker Jr.)

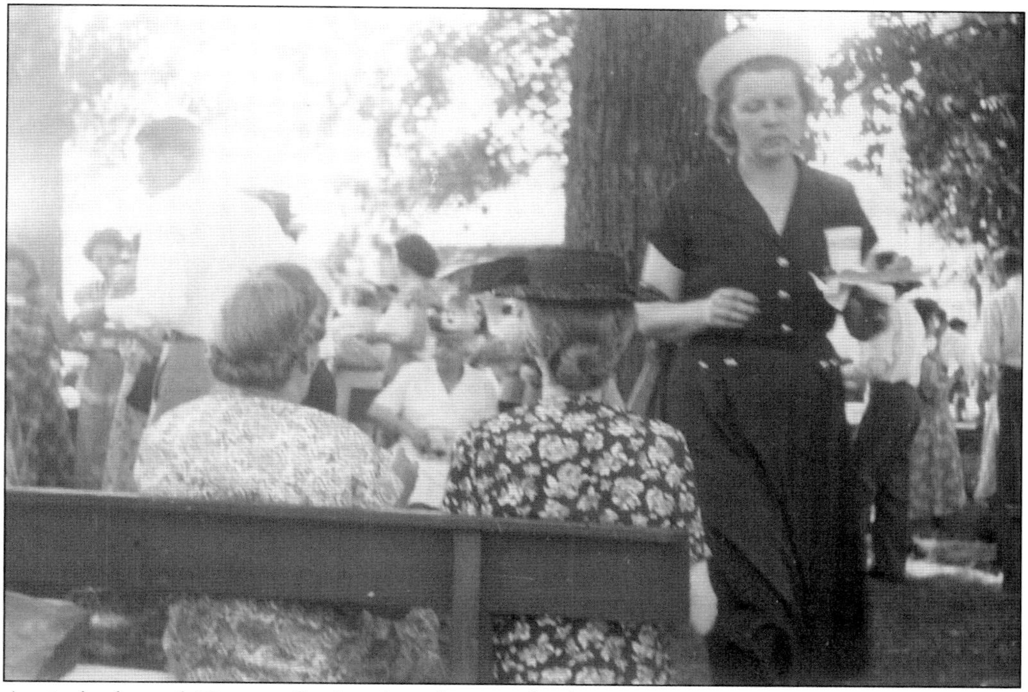

A nicely dressed Virginia Rucker (standing, with plate in hand) is enjoying a Parker family reunion in Moneta in this 1948 photograph. (Courtesy of James Rucker.)

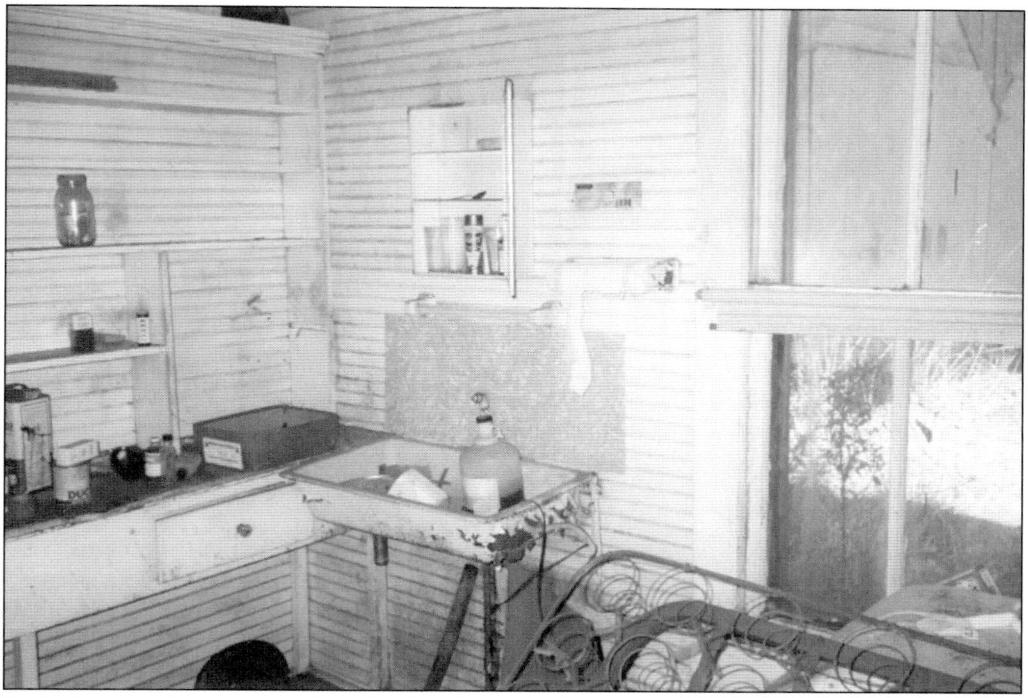

This close-up photograph of Dr. Sam's office, which was taken in the early 1990s, shows the extent of his cramped, humble medical facility. Hardly larger than a tool shed, it was nevertheless adequate for his time. He successfully treated several grateful generations of Bedford County families. (Courtesy of the author.)

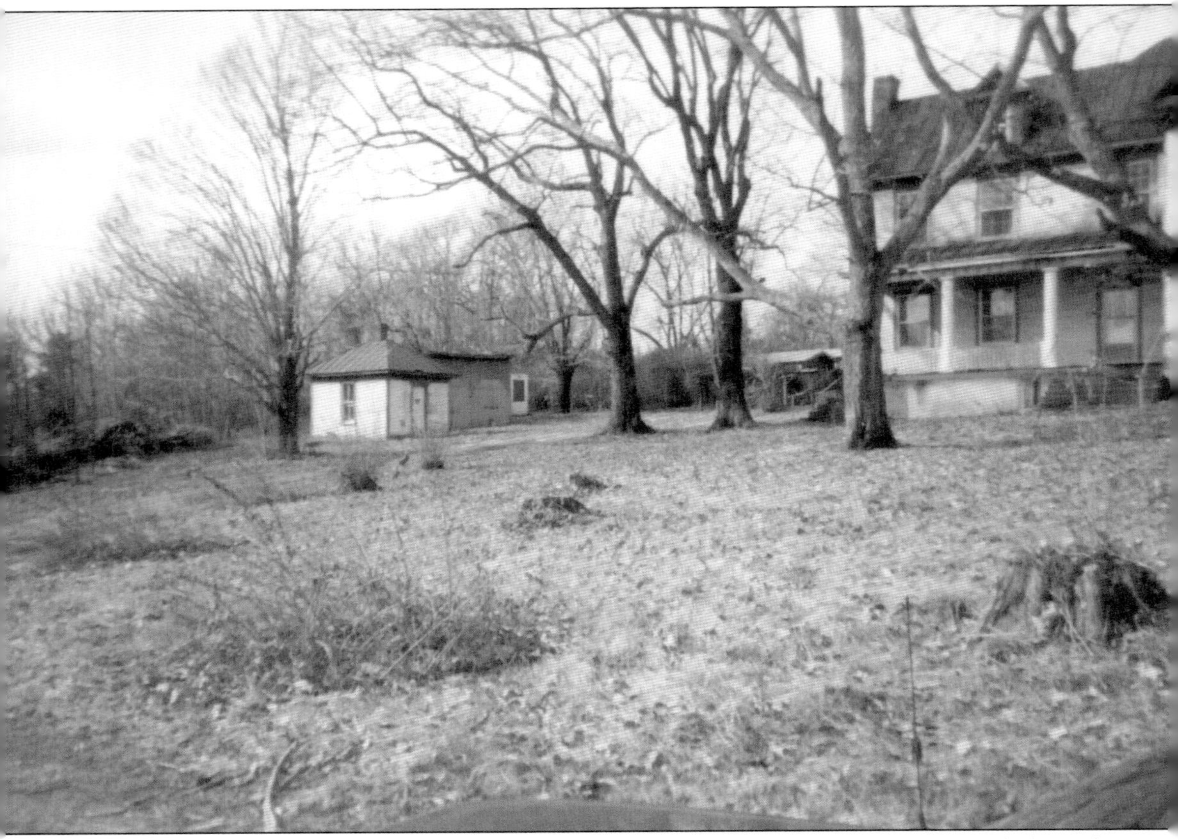

The Sam L. Rucker Jr. House stood empty for a period of time after the doctor's death, but it has been preserved and is used currently for an upscale gift shop. The small, white building on the left is the office he used for his practice until his diabetes prevented the daily walk to and from the house. For years, cars would line up in the parking lot as early as 6:00 a.m. so he could see patients before his daily rounds. At times, various members of the community would accompany him in his automobile when he needed to visit patients. Every five years, the local Ruritan club would replace his heavily mileaged car with a newer model. "Next, if you're sick!" Dr. Sam would bellow from the door of his little office. (Courtesy of the author.)

Five

RIDIN' THE RAILS

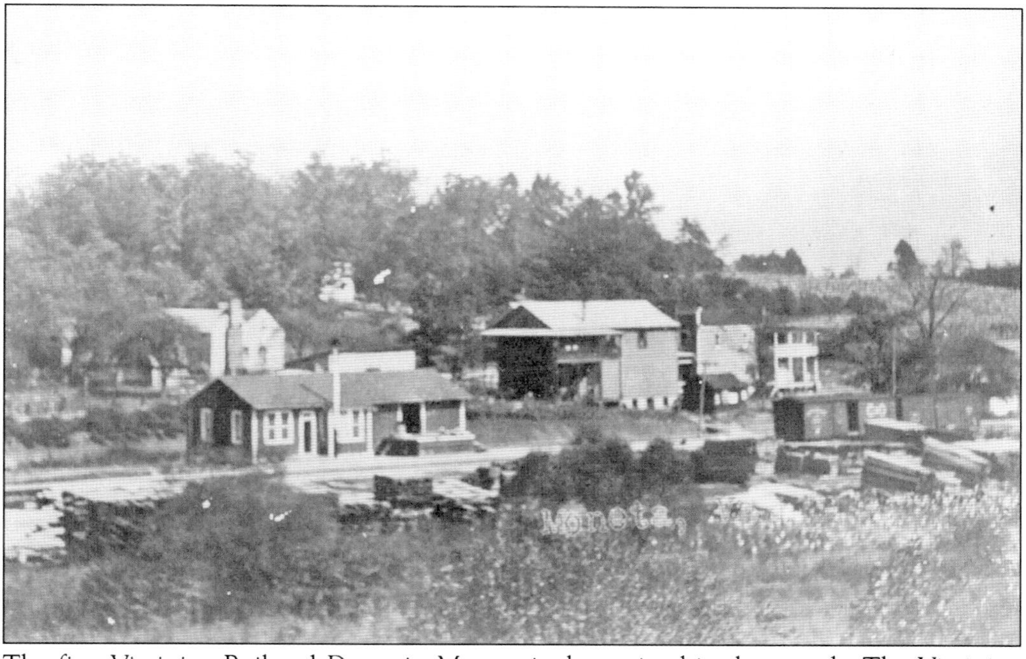

The first Virginian Railroad Depot in Moneta is shown in this photograph. The Virginian Railway merged with the N&W Railroad in 1959. Look to the right and notice the back of the Dave Rucker store and, across the road, the William Howard Meador store. (Courtesy of Gladys Hoover.)

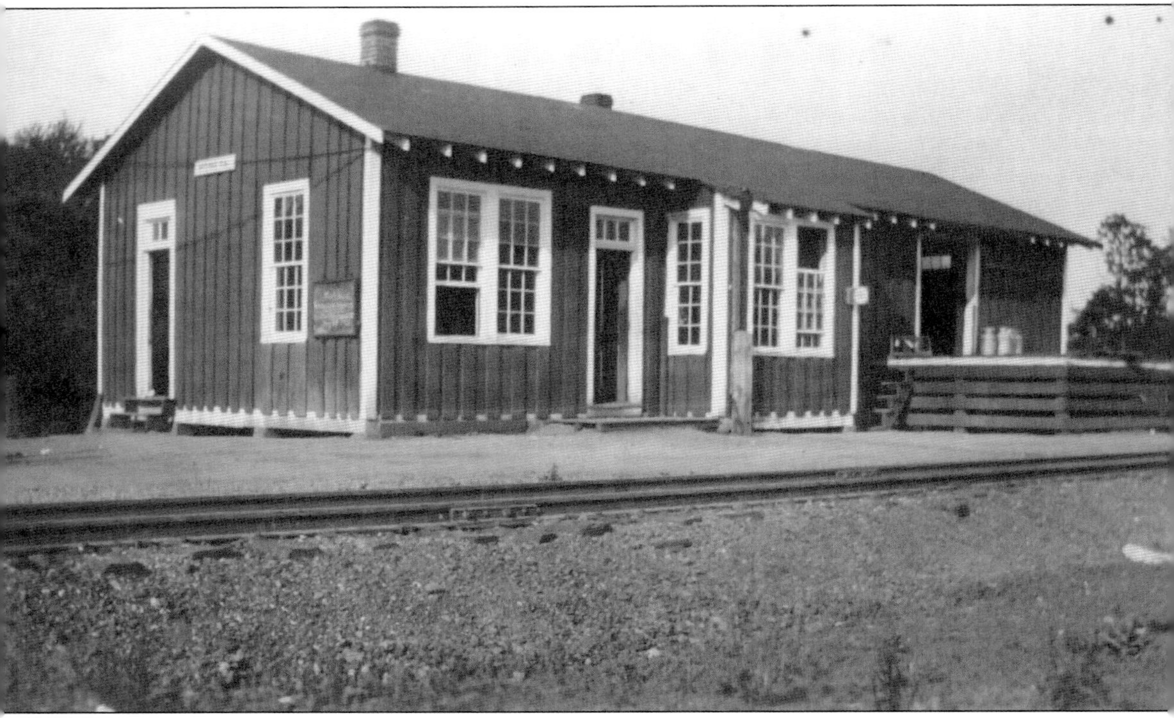

The first Virginian Railroad Depot in Moneta is seen in a closer view in this photograph. (Courtesy of Sam and Debbie Rucker.)

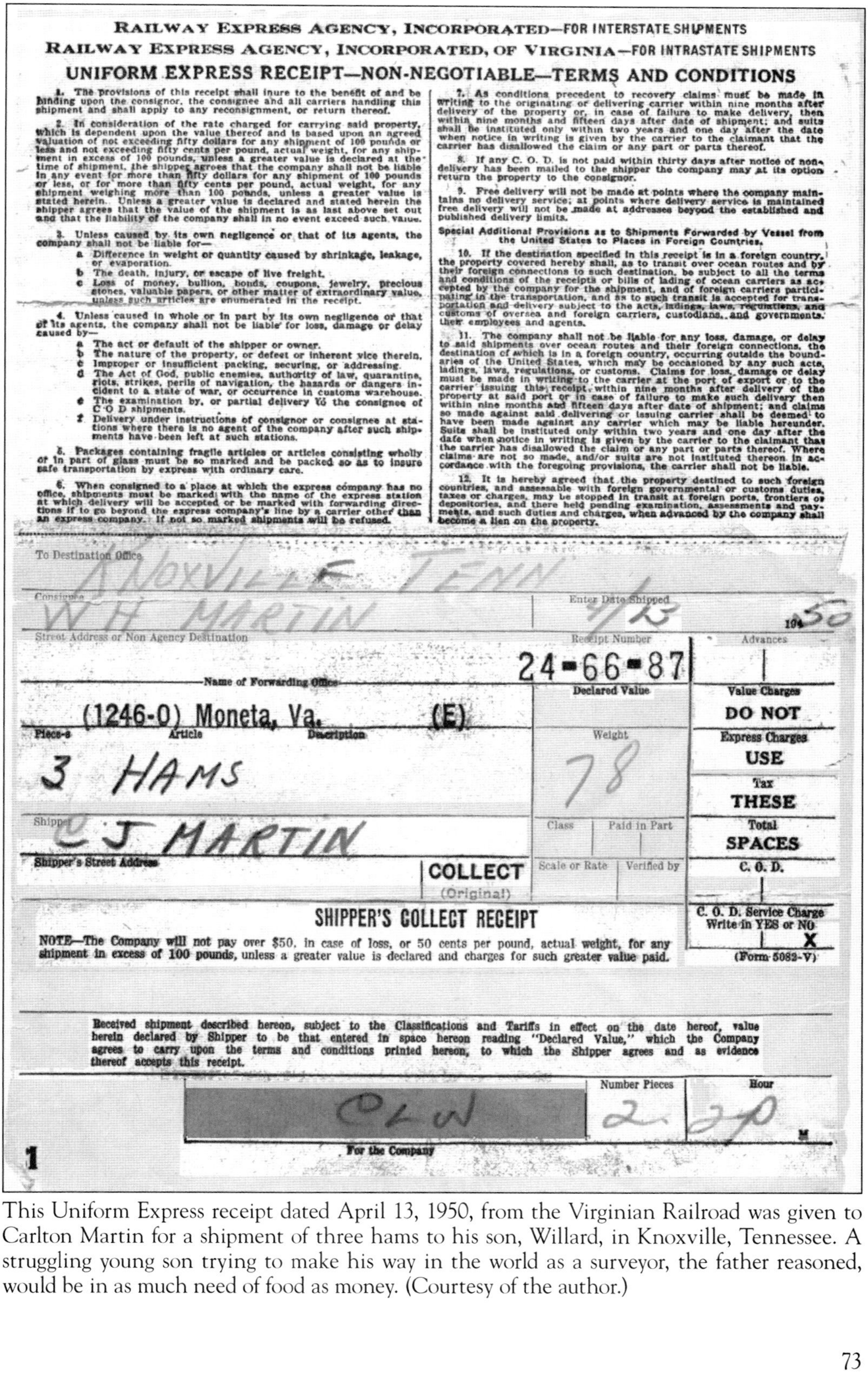

RAILWAY EXPRESS AGENCY, INCORPORATED—FOR INTERSTATE SHIPMENTS

RAILWAY EXPRESS AGENCY, INCORPORATED, OF VIRGINIA—FOR INTRASTATE SHIPMENTS

UNIFORM EXPRESS RECEIPT—NON-NEGOTIABLE—TERMS AND CONDITIONS

1. The provisions of this receipt shall inure to the benefit of and be binding upon the consignor, the consignee and all carriers handling this shipment and shall apply to any reconsignment, or return thereof.

2. In consideration of the rate charged for carrying said property, which is dependent upon the value thereof and is based upon an agreed valuation of not exceeding fifty dollars for any shipment of 100 pounds or less and not exceeding fifty cents per pound, actual weight, for any shipment in excess of 100 pounds, unless a greater value is declared at the time of shipment, the shipper agrees that the company shall not be liable in any event for more than fifty dollars for any shipment of 100 pounds or less, or for more than fifty cents per pound, actual weight, for any shipment weighing more than 100 pounds, unless a greater value is stated herein. Unless a greater value is declared and stated herein the shipper agrees that the value of the shipment is as last above set out and that the liability of the company shall in no event exceed such value.

3. Unless caused by its own negligence or that of its agents, the company shall not be liable for—

a Difference in weight or quantity caused by shrinkage, leakage, or evaporation.
b The death, injury, or escape of live freight.
c Loss of money, bullion, bonds, coupons, jewelry, precious stones, valuable papers, or other matter of extraordinary value, unless such articles are enumerated in the receipt.

4. Unless caused in whole or in part by its own negligence or that of its agents, the company shall not be liable for loss, damage or delay caused by—

a The act or default of the shipper or owner.
b The nature of the property, or defect or inherent vice therein.
c Improper or insufficient packing, securing, or addressing.
d The Act of God, public enemies, authority of law, quarantine, riots, strikes, perils of navigation, the hazards or dangers incident to a state of war, or occurrence in customs warehouse.
e The examination by, or partial delivery to the consignee of C O D shipments.
f Delivery under instructions of consignor or consignee at stations where there is no agent of the company after such shipments have been left at such stations.

5. Packages containing fragile articles or articles consisting wholly or in part of glass must be so marked and be packed so as to insure safe transportation by express with ordinary care.

6. When consigned to a place at which the express company has no office, shipments must be marked with the name of the express station at which delivery will be accepted or be marked with forwarding directions if to go beyond the express company's line by a carrier other than an express company. If not so marked shipments will be refused.

7. As conditions precedent to recovery claims must be made in writing to the originating or delivering carrier within nine months after delivery of the property or, in case of failure to make delivery, then within nine months and fifteen days after date of shipment; and suits shall be instituted only within two years and one day after the date when notice in writing is given by the carrier to the claimant that the carrier has disallowed the claim or any part or parts thereof.

8. If any C. O. D. is not paid within thirty days after notice of non-delivery has been mailed to the shipper the company may at its option return the property to the consignor.

9. Free delivery will not be made at points where the company maintains no delivery service; at points where delivery service is maintained free delivery will not be made at addresses beyond the established and published delivery limits.

Special Additional Provisions as to Shipments Forwarded by Vessel from the United States to Places in Foreign Countries.

10. If the destination specified in this receipt is in a foreign country, the property covered hereby shall, as to transit over ocean routes and by their foreign connections to such destination, be subject to all the terms and conditions of the receipts or bills of lading of ocean carriers as accepted by the company for the shipment, and of foreign carriers participating in the transportation, and as to such transit is accepted for transportation and delivery subject to the acts, ladings, laws, regulations, and customs of oversea and foreign carriers, custodians, and governments, their employees and agents.

11. The company shall not be liable for any loss, damage, or delay to said shipments over ocean routes and their foreign connections, the destination of which is in a foreign country, occurring outside the boundaries of the United States, which may be occasioned by any such acts, ladings, laws, regulations, or customs. Claims for loss, damage or delay must be made in writing to the carrier at the port of export or to the carrier issuing this receipt, within nine months after delivery of the property at said port or in case of failure to make such delivery then within nine months and fifteen days after date of shipment; and claims so made against said delivering or issuing carrier shall be deemed to have been made against any carrier which may be liable hereunder. Suits shall be instituted only within two years and one day after the date when notice in writing is given by the carrier to the claimant that the carrier has disallowed the claim or any part or parts thereof. Where claims are not so made, and/or suits are not instituted thereon in accordance with the foregoing provisions, the carrier shall not be liable.

12. It is hereby agreed that the property destined to such foreign countries, and assessable with foreign governmental or customs duties, taxes or charges, may be stopped in transit at foreign ports, frontiers or depositories, and there held pending examination, assessments and payments, and such duties and charges, when advanced by the company shall become a lien on the property.

To Destination Office: KNOXVILLE TENN

Consignee: W H MARTIN

Enter Date Shipped: 4/13 1950

Street Address or Non Agency Destination

Receipt Number: 24-66-87

Advances

Name of Forwarding Office: (1246-0) Moneta, Va. (E)

Declared Value	Value Charges
	DO NOT
Weight: 78	Express Charges: USE
	Tax: THESE
Class / Paid in Part	Total: SPACES
Scale or Rate / Verified by	C. O. D.

Pieces / Article / Description: 3 HAMS

Shipper: C J MARTIN

Shipper's Street Address

COLLECT (Original)

SHIPPER'S COLLECT RECEIPT

C. O. D. Service Charge Write in YES or NO: X

NOTE—The Company will not pay over $50, in case of loss, or 50 cents per pound, actual weight, for any shipment in excess of 100 pounds, unless a greater value is declared and charges for such greater value paid.

(Form 5082-V)

Received shipment described hereon, subject to the Classifications and Tariffs in effect on the date hereof, value herein declared by Shipper to be that entered in space hereon reading "Declared Value," which the Company agrees to carry upon the terms and conditions printed hereon, to which the Shipper agrees and as evidence thereof accepts this receipt.

For the Company	Number Pieces	Hour
[illegible]	2	[illegible] M

1

This Uniform Express receipt dated April 13, 1950, from the Virginian Railroad was given to Carlton Martin for a shipment of three hams to his son, Willard, in Knoxville, Tennessee. A struggling young son trying to make his way in the world as a surveyor, the father reasoned, would be in as much need of food as money. (Courtesy of the author.)

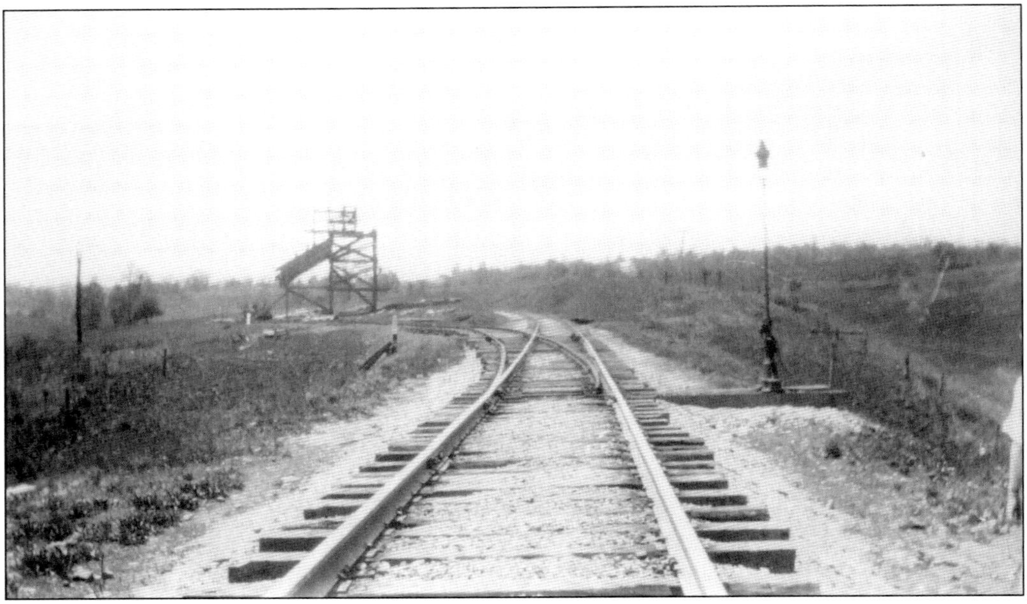

Mary Rucker's caption on this May 8, 1924, photograph reads, " 'Switch and Tipple,' Moneta Mineral and Mining Company." (Courtesy of Sam and Debbie Rucker.)

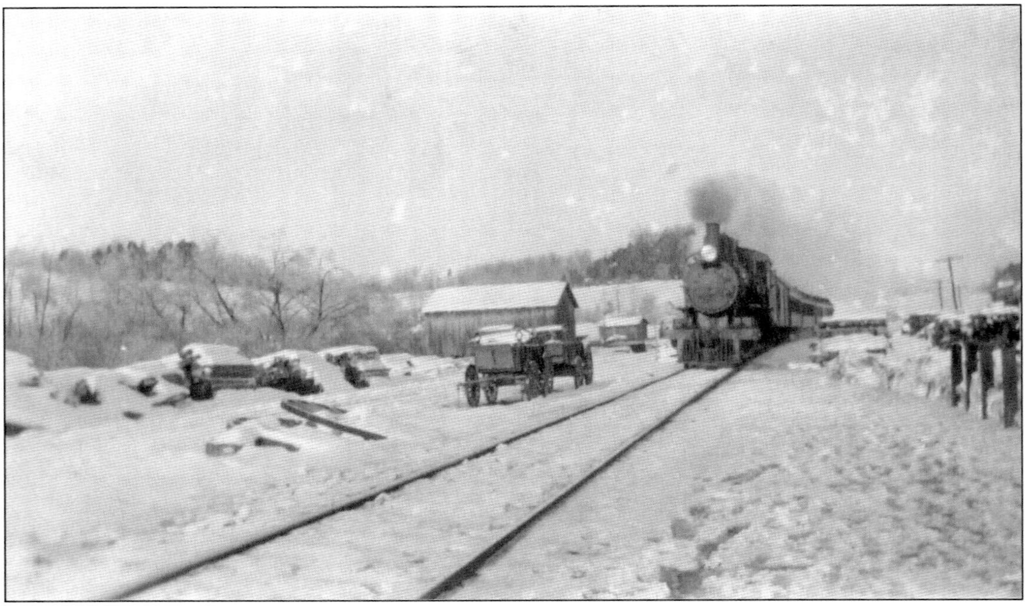

The caption of this photograph, taken by Mary Rucker, reads, "Train No. 2 at Moneta, Virginia." (Courtesy of Sam and Debbie Rucker.)

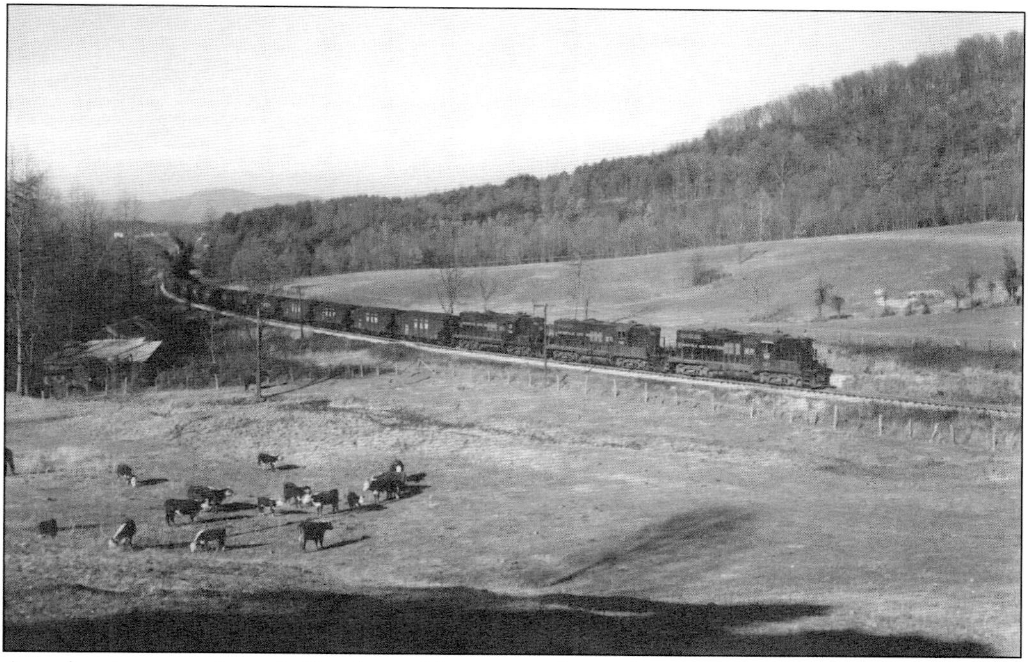

A coal train passes through Goodview, Virginia, in this undated photograph. (NS3526, courtesy of the Norfolk and Western Historical Photograph Collection and the Digital Library and Archives, University Libraries, Virginia Polytechnic Institute and State University.)

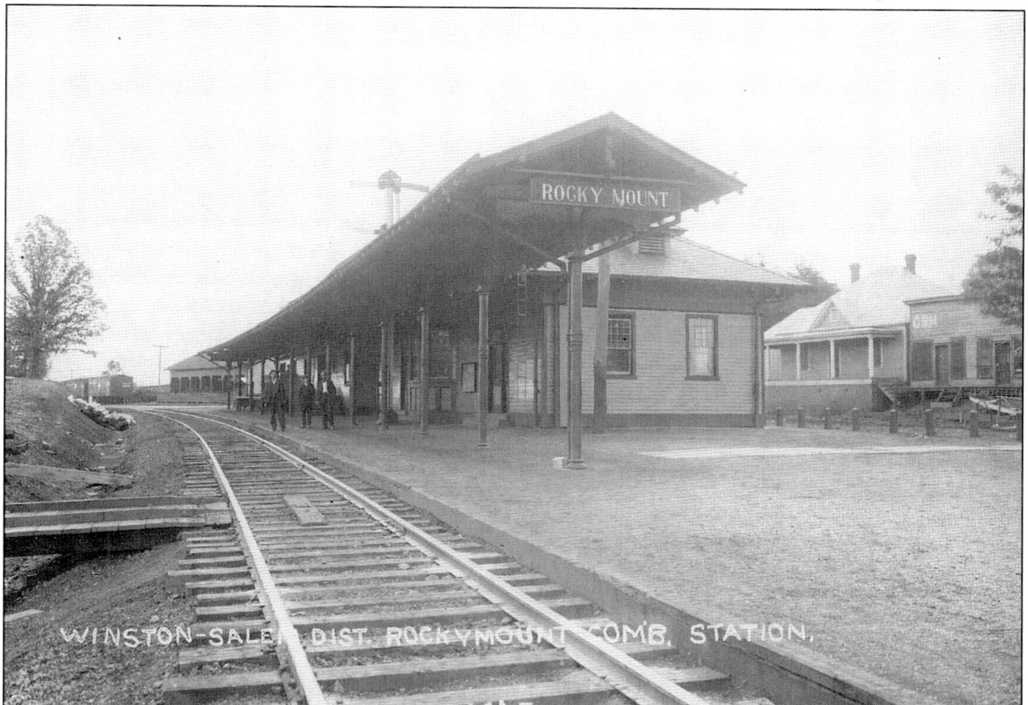

This was the Rocky Mount, Virginia, combination station for both passengers and freight. (NS490, courtesy of the Norfolk and Western Historical Photograph Collection and the Digital Library and Archives, University Libraries, Virginia Polytechnic Institute and State University.)

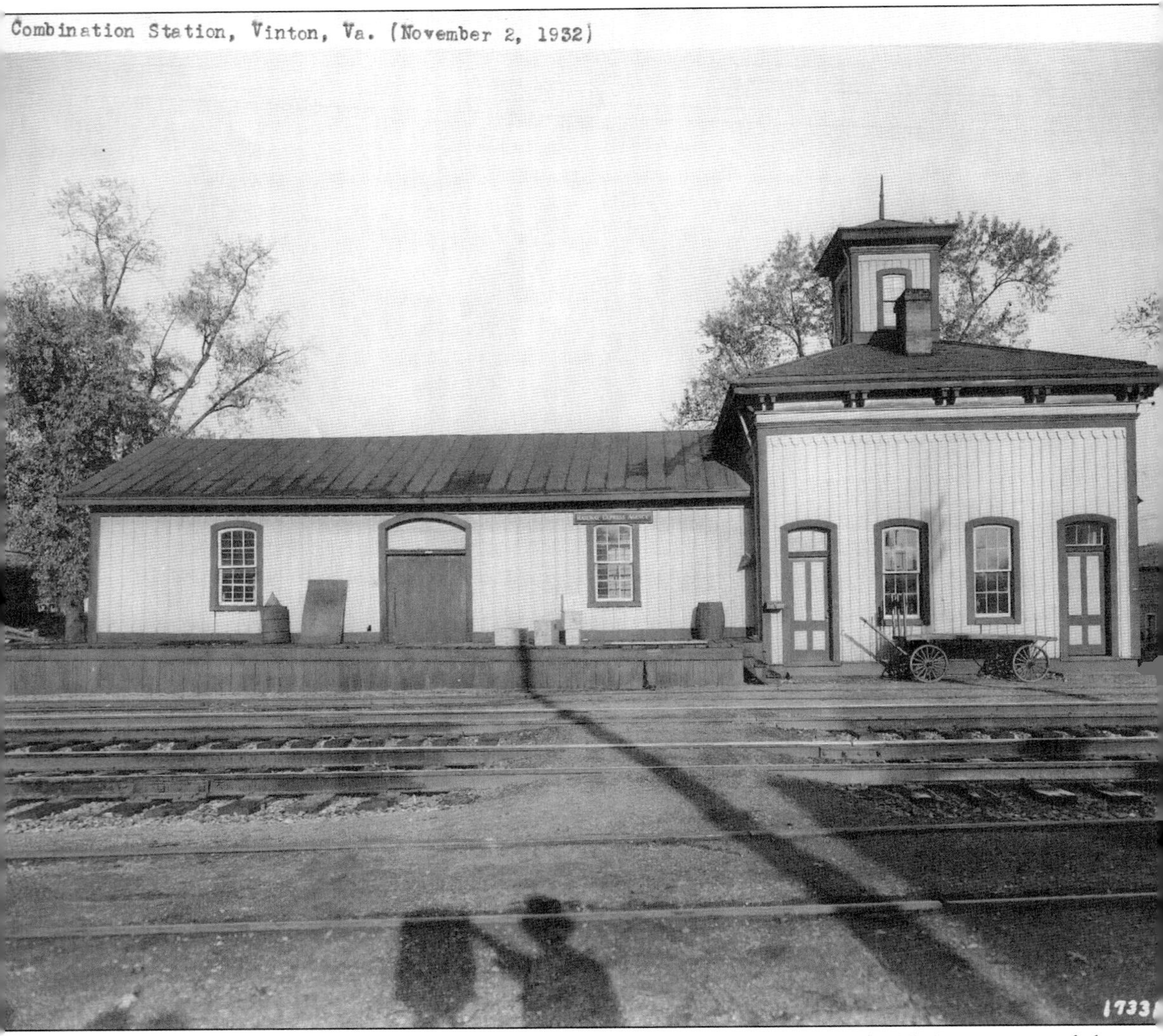

This photograph shows the combination station in Vinton, Virginia. (NS542, courtesy of the Norfolk and Western Historical Photograph Collection and the Digital Library and Archives, University Libraries, Virginia Polytechnic Institute and State University.)

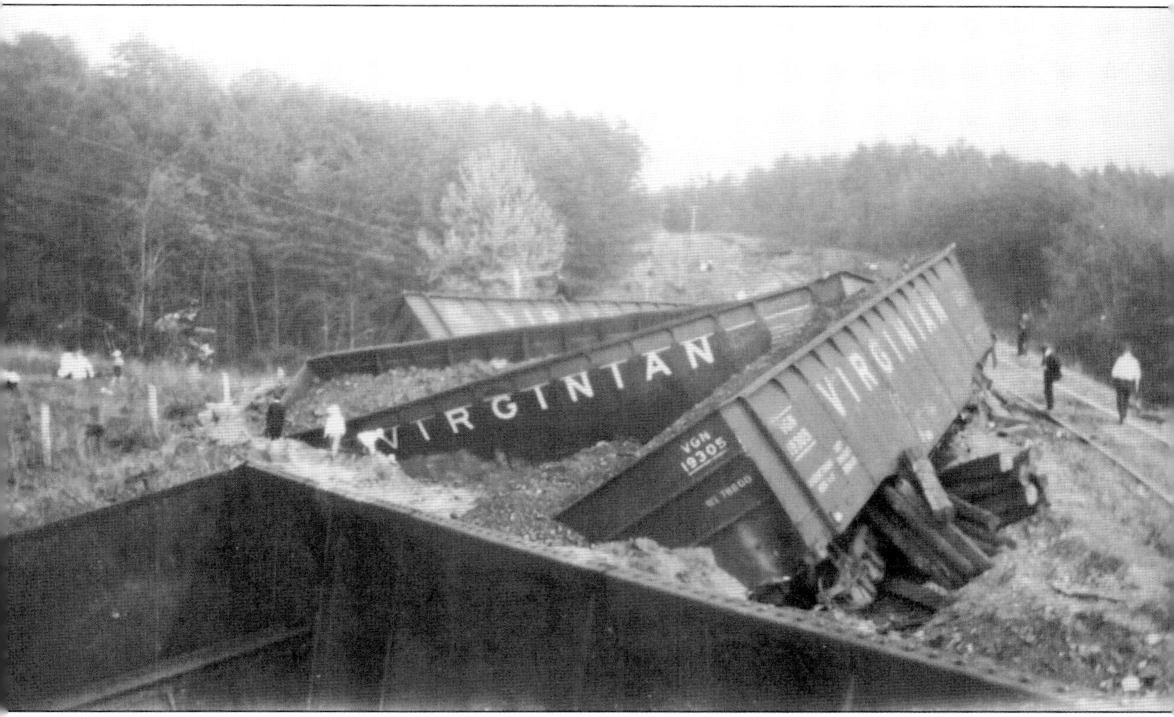

This photograph was taken one day after a May 5, 1928, wreck of a Virginian Railroad train near Morgans Baptist Church in Moneta. (Courtesy of Sam and Debbie Rucker.)

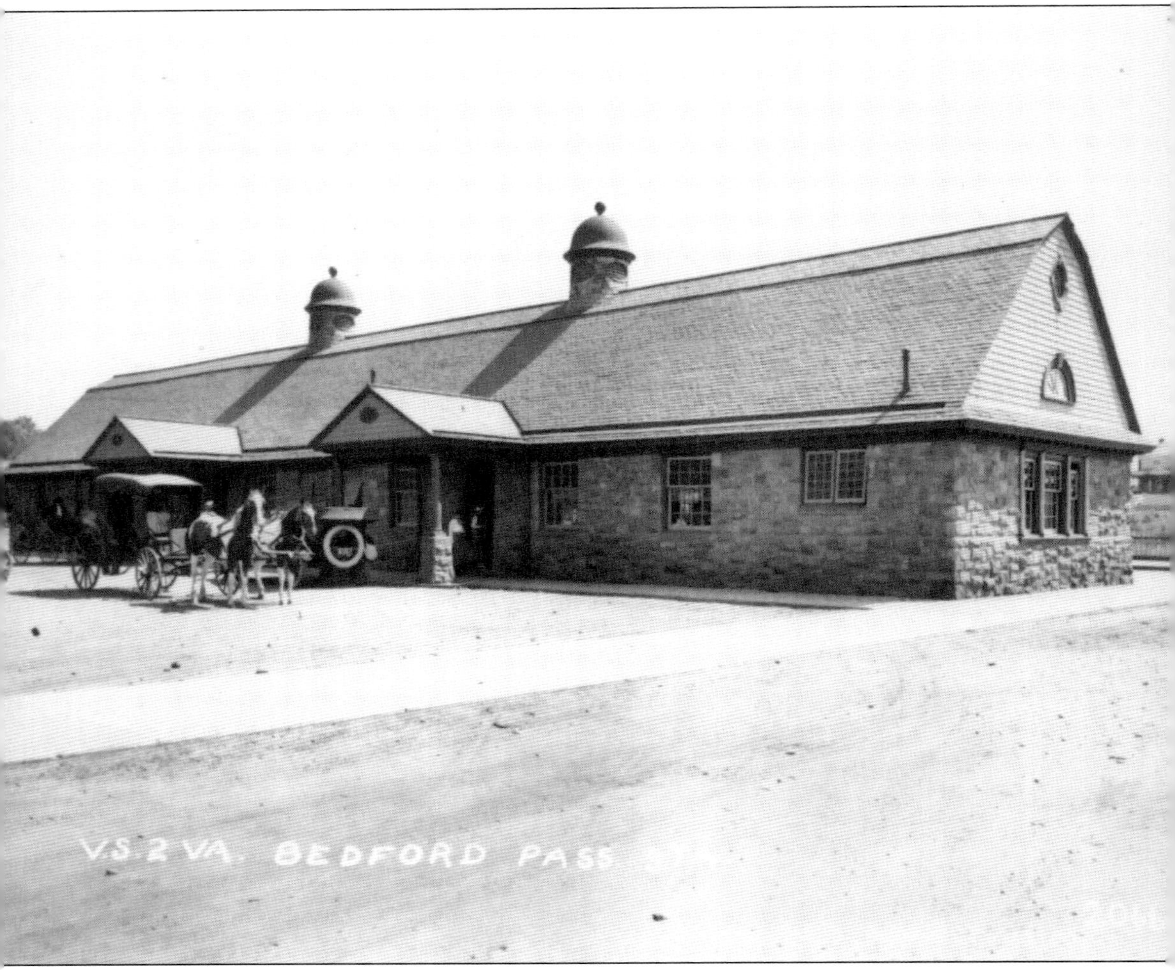

The Bedford train station was built in 1905. After serving the community for decades as the transporter of passengers, freight, and mail, it would become a restaurant in later years. It was renamed Olde Liberty Station in 1997 by its owners. (Courtesy of Frank Ewald.)

Six

School Days

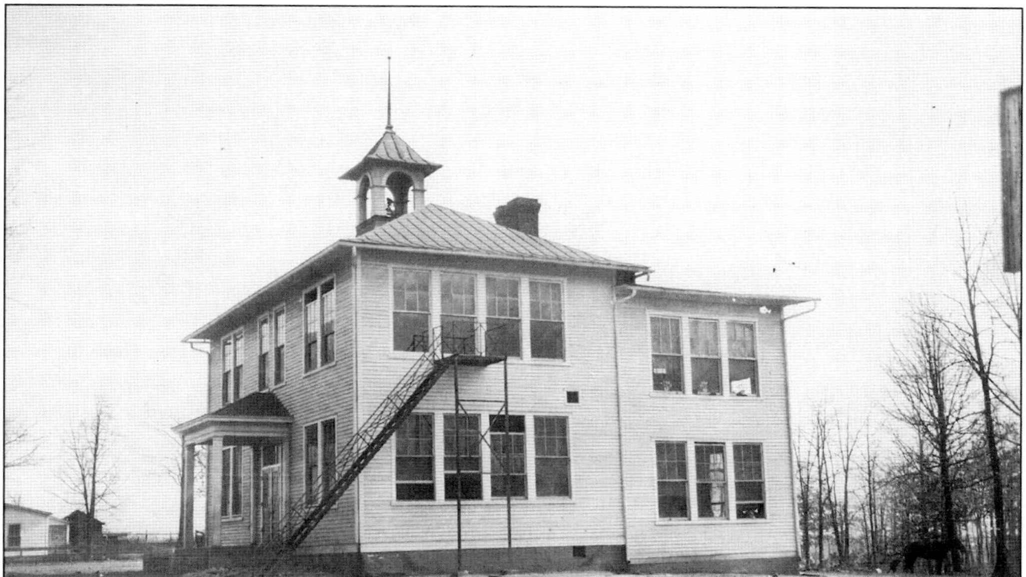

This is a front view of the Moneta High School, which stood on the grounds of the current Moneta Elementary School, in 1930. It was built in 1909 and was in operation until a new brick high school was built on land behind it in 1934. Though the old school was torn down, the house in the background (bottom left) is still standing. (Courtesy of the Library of Virginia.)

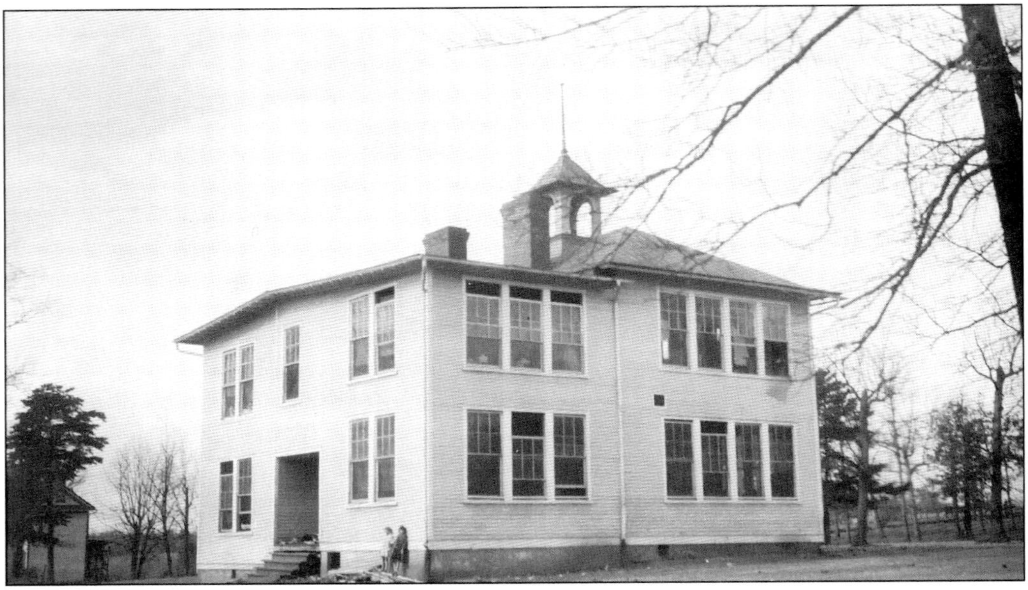

In this view of the back of Moneta School, two youngsters appear to be hiding or skipping classes. This view, which faces the village of Moneta, shows an obvious addition to the original structure that was added around 1916. (Courtesy of the Library of Virginia.)

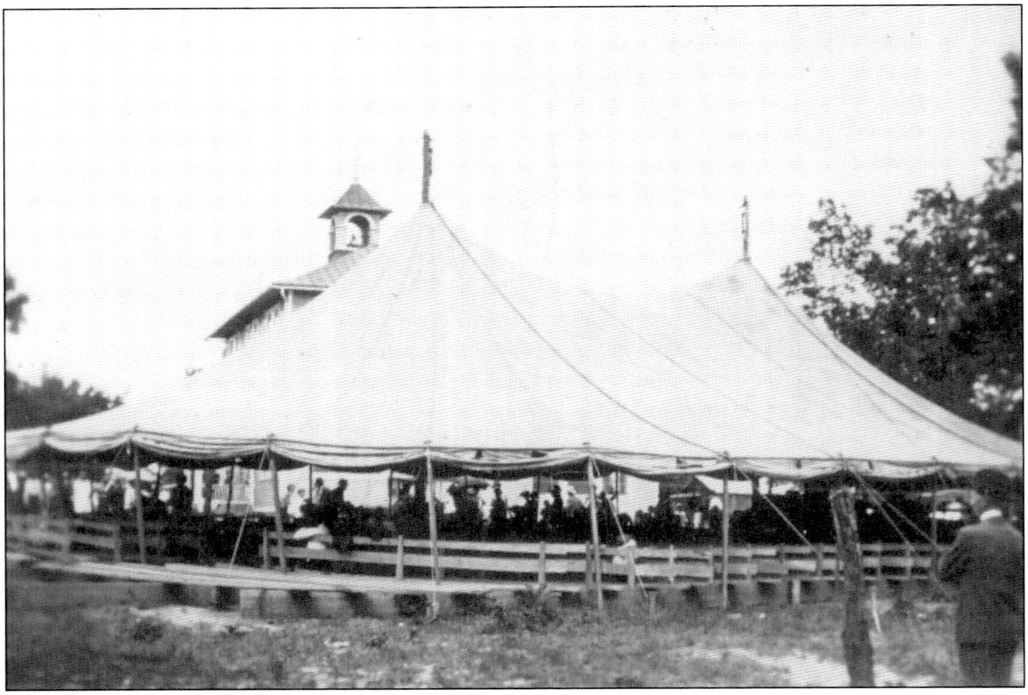

From July 1 to July 15, 1923, this tent stood on the grounds of the first Moneta High School, according to Mary Rucker's caption. (Courtesy of Sam and Debbie Rucker.)

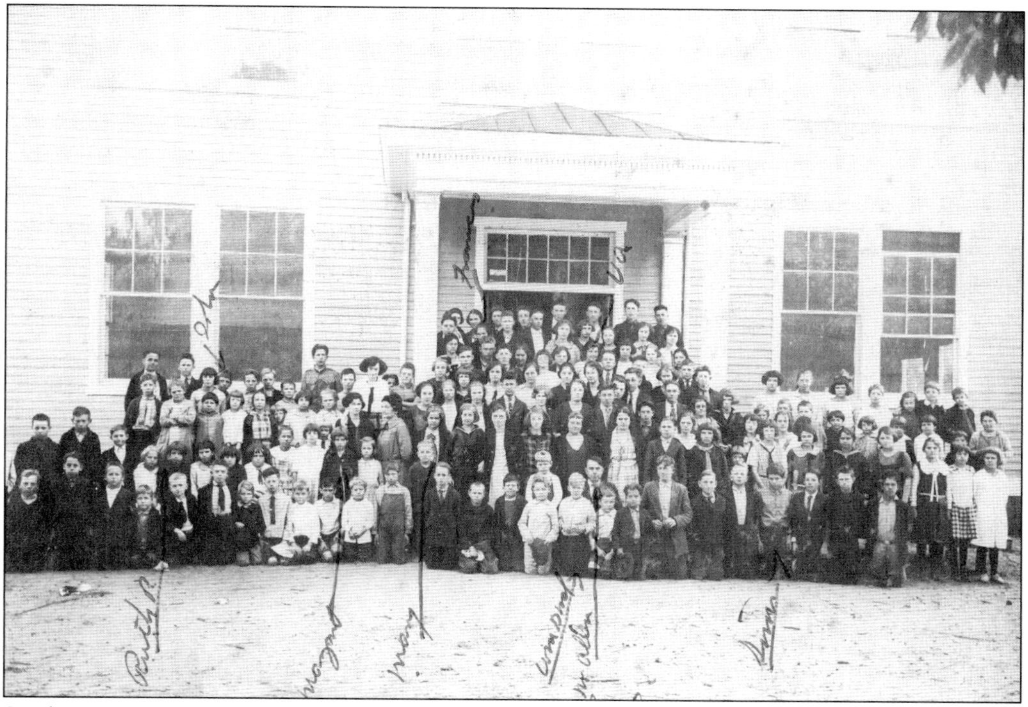

Students of Moneta High School pose for a spring 1923 photograph. (Courtesy of Dr. James Rucker Jr.)

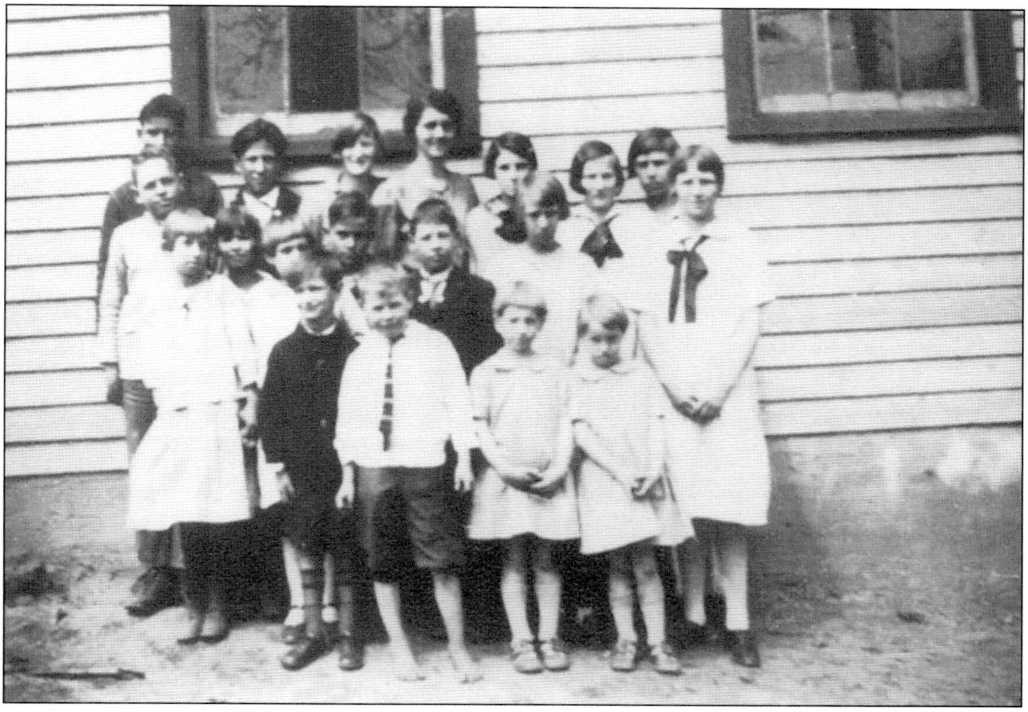

Students of the Western Light School pose outside in this 1930s photograph. The school was located on State Route 122 before being moved to Hendricks Store Road in Moneta. (Courtesy of Shirley Ring.)

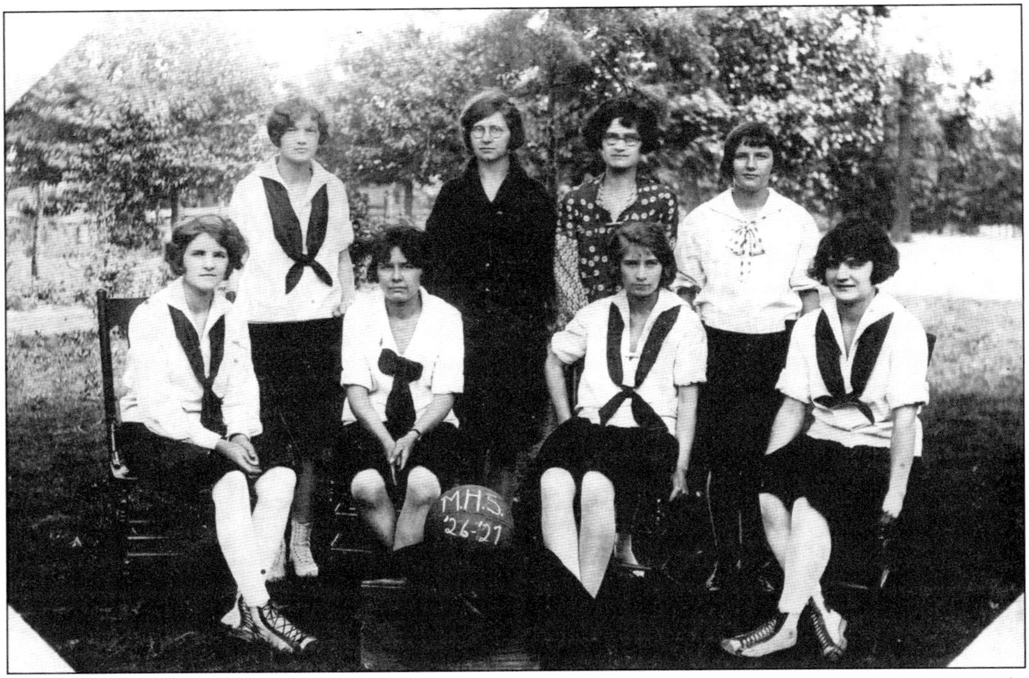

The 1926–1927 Moneta High School girls' basketball team poses with a dated basketball. Only two of the players could be identified: on the left, seated, is Lucy Kate Bradley; third from the left, seated, is Elizabeth Meador, mother of Gladys Hoover. (Courtesy of Gladys Hoover.)

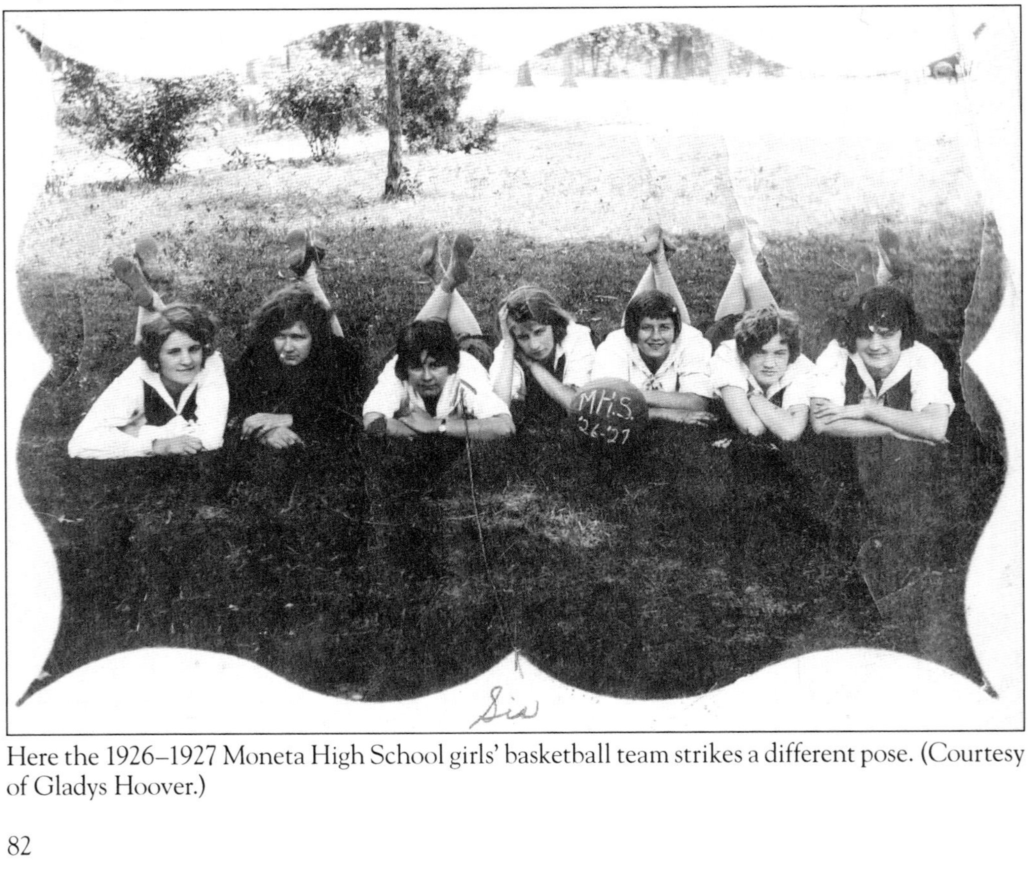

Here the 1926–1927 Moneta High School girls' basketball team strikes a different pose. (Courtesy of Gladys Hoover.)

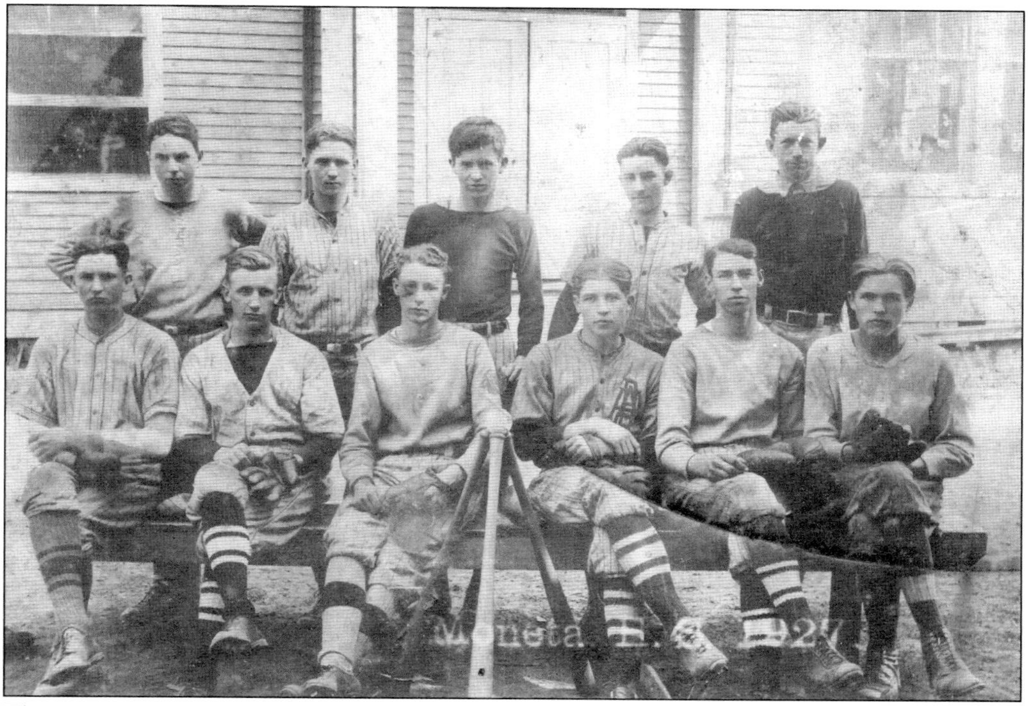

The 1927 Moneta High School boys' baseball team poses for a picture. (Courtesy of Gladys Hoover.)

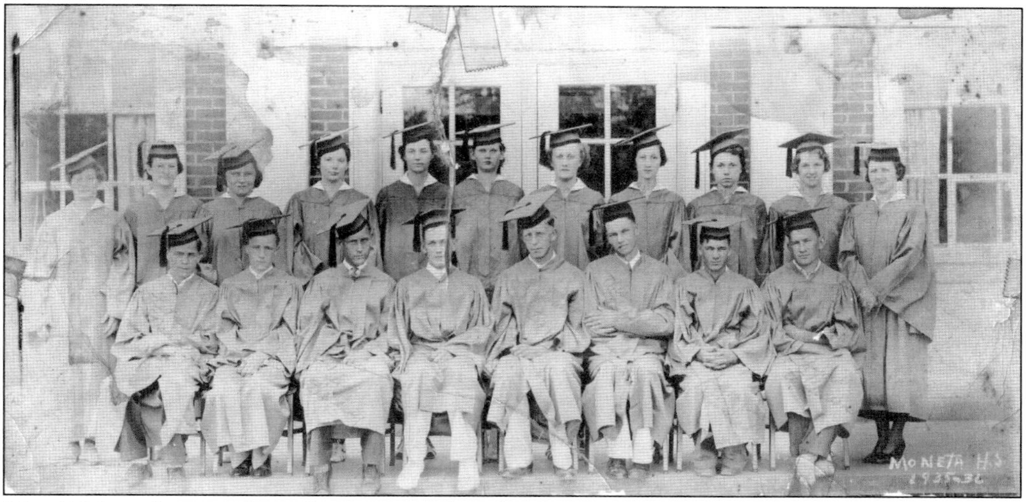

The 1936 graduating class of Moneta High School poses for a formal photograph. (Courtesy of Gladys Hoover.)

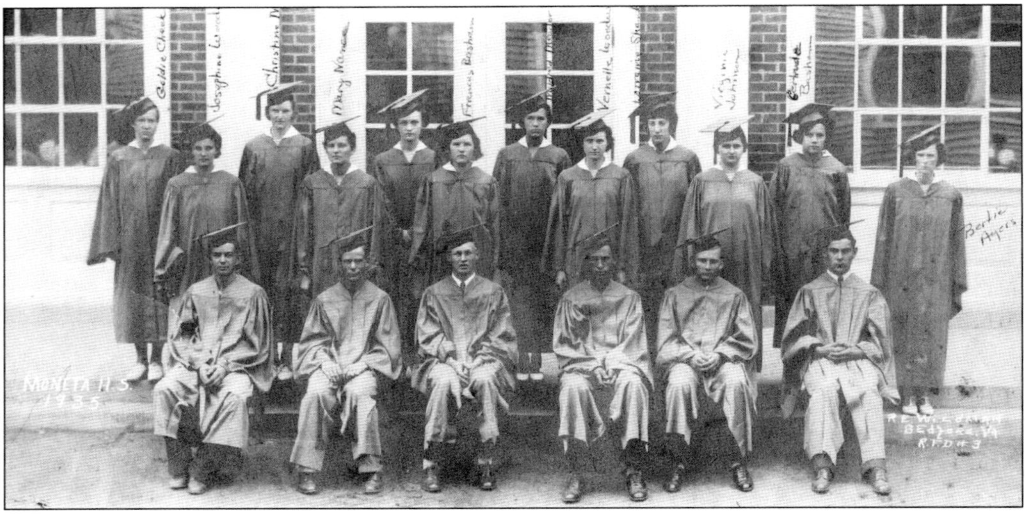

The graduating class of 1935 posed for this cap and gown photograph. They are, from left to right, (front row) Richard Rucker, Watson Nance, Lyle Morgan, Bricen Saunders, Ted Woodward, and Malcolm Bowyer; (second row) Goldie Cheek, Josephine Woodward, Christine Nance, Mary Nance, Laura Morgan, Frances Basham, Mildred Meador, Vernelle Woodward, Lorraine Shank, Virginia Johnson, Gertrude Basham, and Bertie Ayers. (Courtesy of Gladys Hoover.)

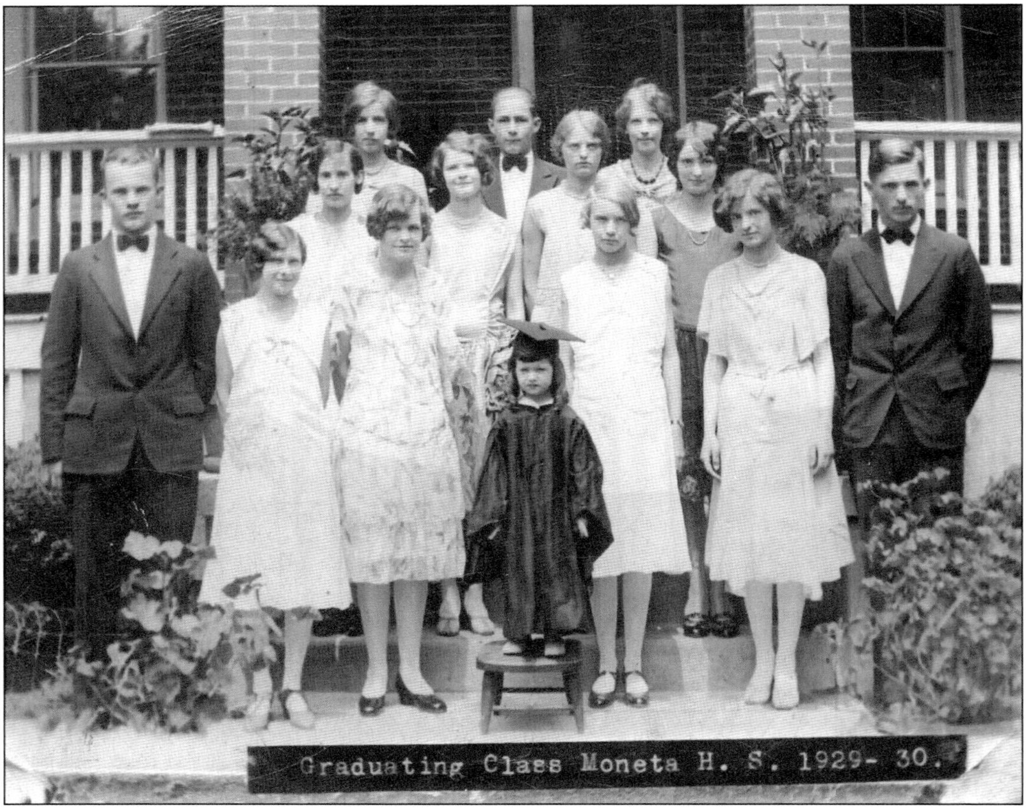

The 1930 graduating class of Moneta High School poses outside the Dave Rucker home. The home was located across from the current Moneta Elementary School, on the same site as the current Moneta Fire and Rescue Department. (Courtesy of Gladys Hoover.)

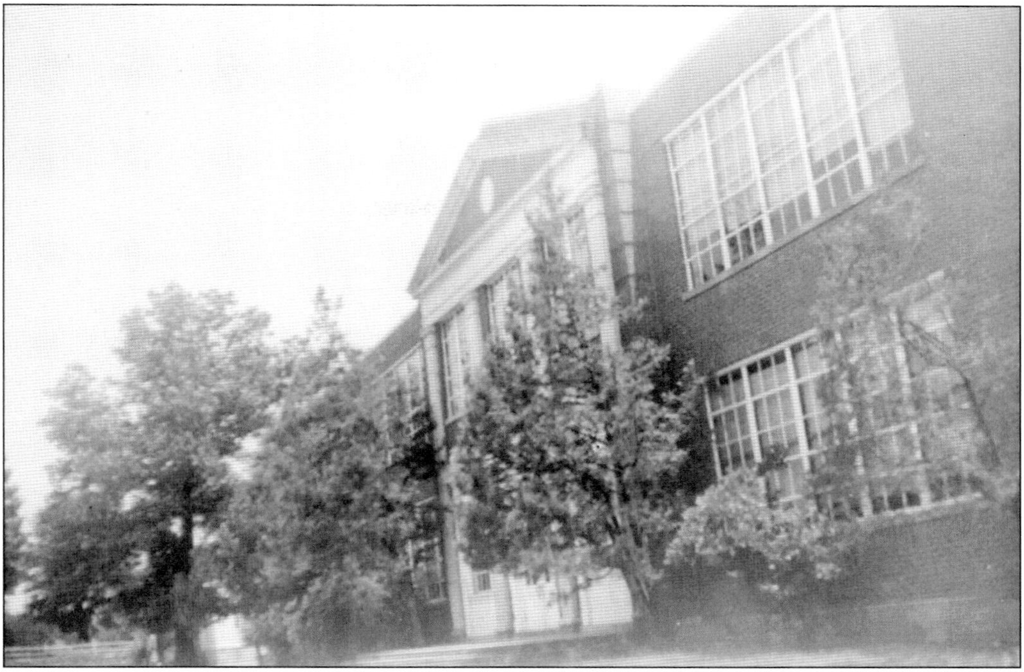

The second Moneta High School stood on the location of the first high school and the current Moneta Elementary School. It was in operation from 1935 to 1963, the year when Huddleston, Stewartsville, and Moneta Highs were merged to form Staunton River High School. (Courtesy of Shirley Ring.)

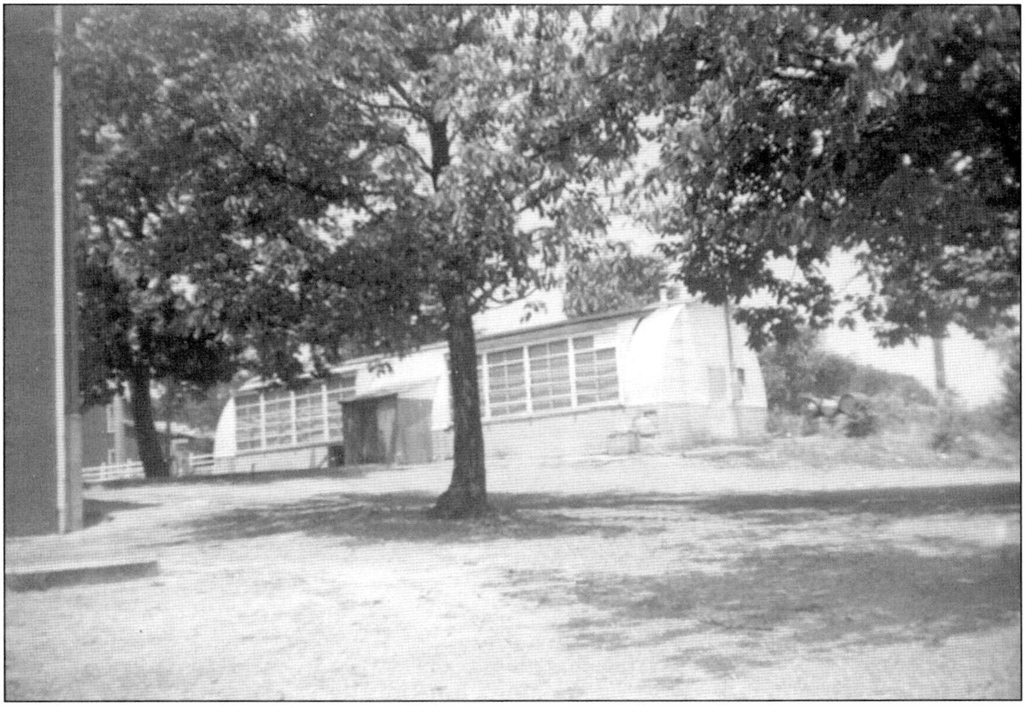

The cafeteria of the second Moneta High School was detached from and stood behind the main facility. (Courtesy of Shirley Ring.)

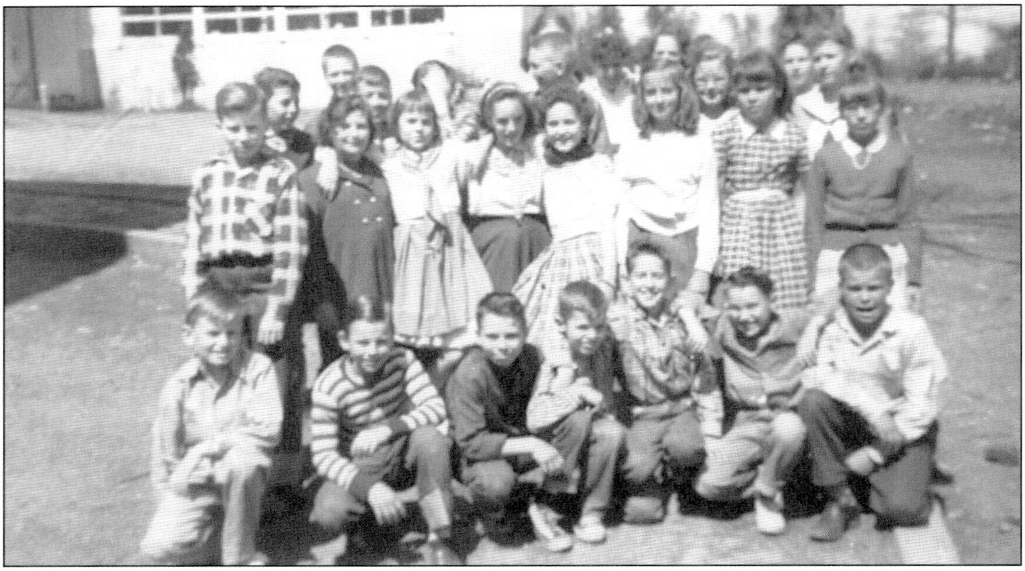

The fifth-grade class of Moneta School posed in front of the cafeteria in this 1956 photograph. (Courtesy of Shirley Ring.)

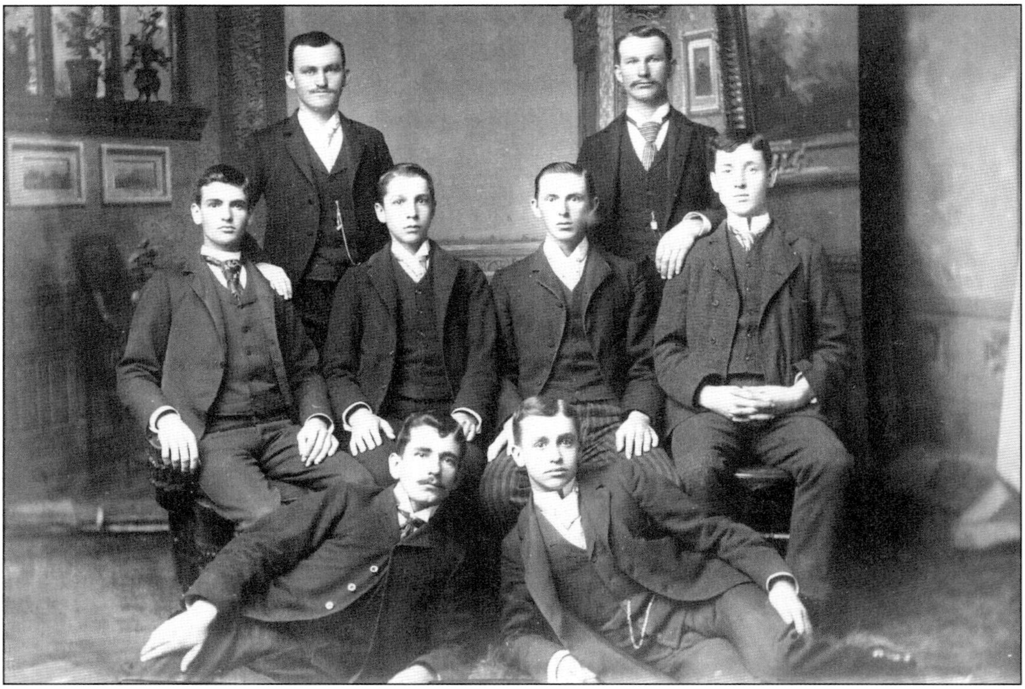

The Bachelor Boys pose for a May 1, 1891, photograph at the New London Academy. On the second row, third from the left, is Dr. Sam L. Rucker Sr. According to the New London Museum Web site (www.newlondonmuseum.org), "Thomas Jefferson's vision for the education of the masses of the people inspired concerned citizens of this area to request from the General Assembly of Virginia a charter 'to incorporate trustees to an academy in the vicinity of New London.' On December 1, 1795, the Virginia General Assembly approved this bill and appointed prominent Bedford County men as trustees. Thus, New London Academy, named for the historic little village had its beginning." (Courtesy of Dr. James Rucker Jr.)

According to *Bedford Villages Lost and Found* Volume 3, the Western Light School was moved from its prior location on State Route 122 (near the present-day Resurrection Catholic Church) to Hendricks Store Road. The building, which is still standing though no longer in use, was renamed Promise Land and "served as a school for black children during segregation." (Courtesy of Shirley Ring.)

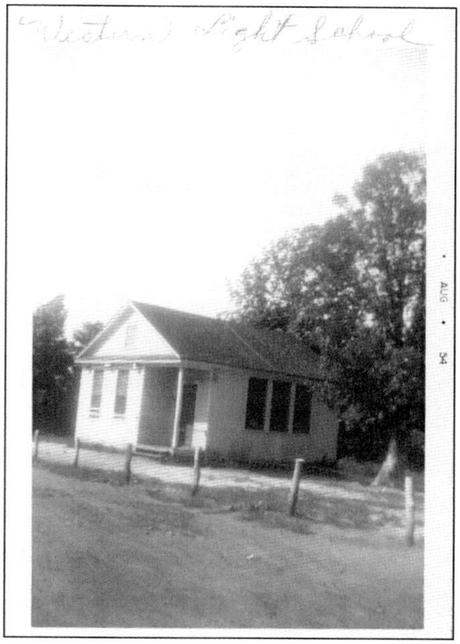

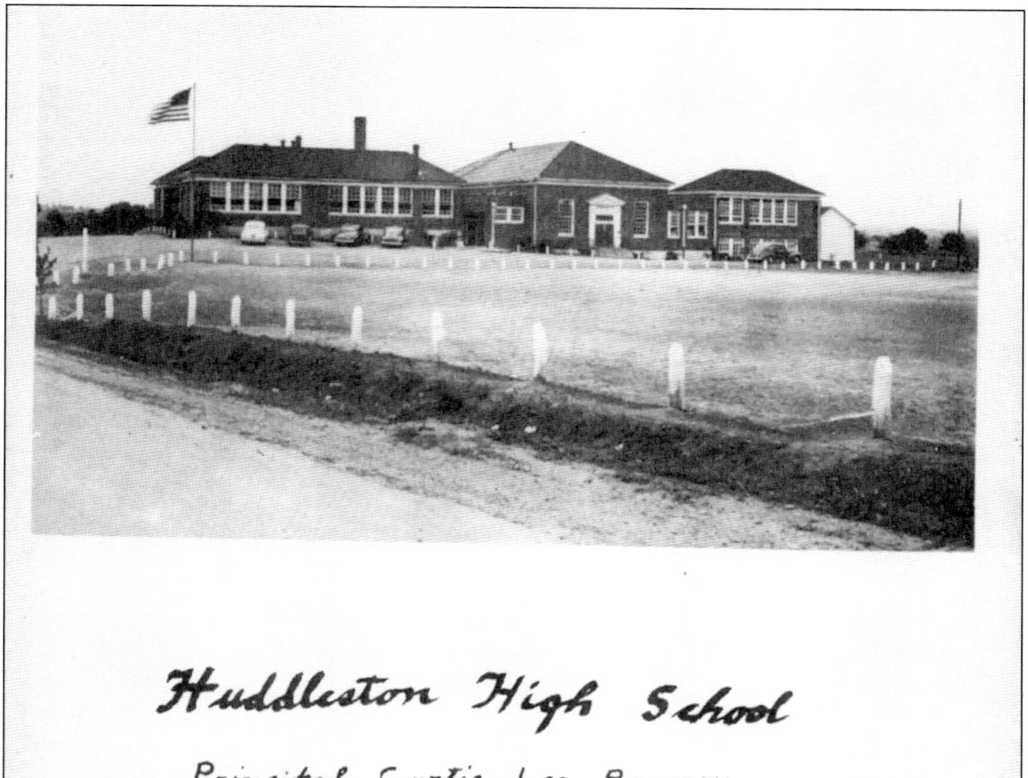

This is a 1940s photograph of what was once Huddleston High School in Huddleston, Virginia. As of May 2008, classmen who attended the institution as early as 1942 are attempting to reconnect with their fellow school chums on the networking site Classmates.com. (Courtesy of the author.)

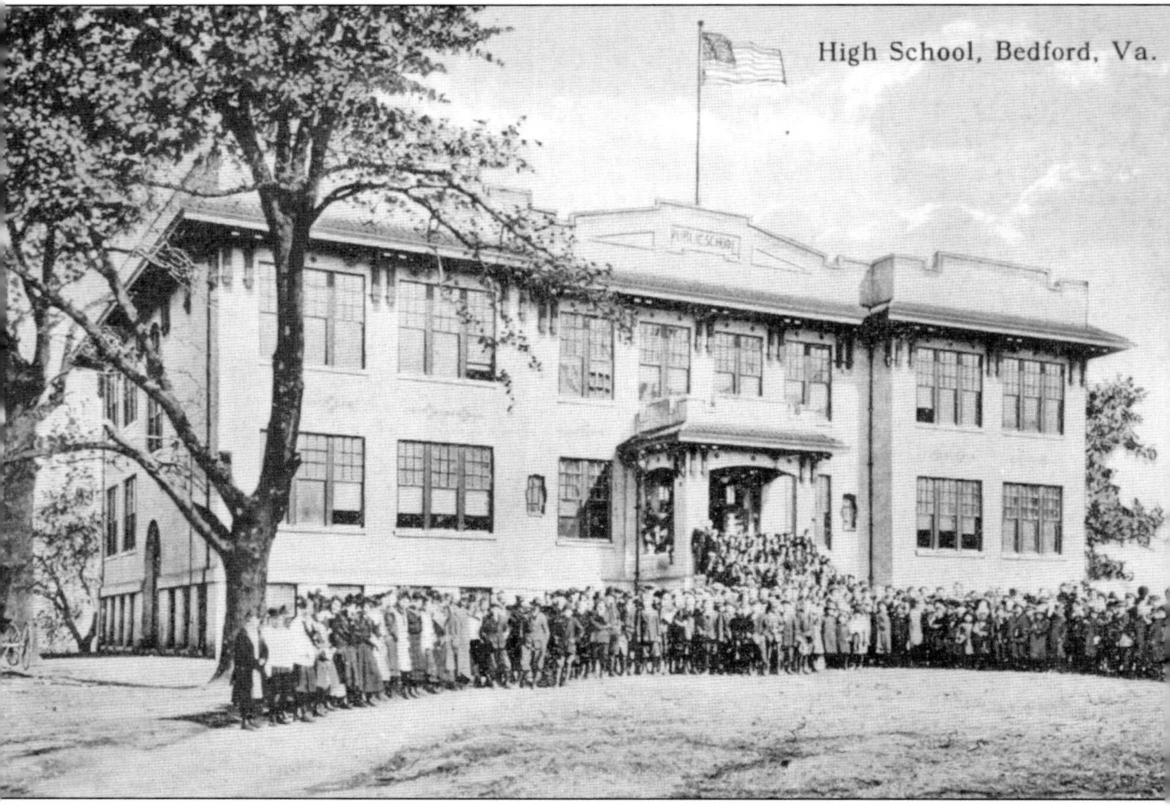

The Bedford High School on Bridge Street and its student body are captured on the back of this undated postcard. Beginning in 1964, students would study at the new Liberty High School located several miles from this structure. (Courtesy of Dr. James Rucker Jr.)

Upon completion of her high school studies at Patmos Normal School in Huddleston, Fawnie Huffman attended two six-week summer sessions at Radford State Teacher's College (now Radford University) and earned a certificate to teach the first grade. This contract with the state board of education stipulated that she would teach at Black's School, a one-room school in Bedford County, for a seven-month period beginning September 25, 1922, at a rate of $60 per month. At the bottom of the contract, note that the school board agreed to pay $1.50 per month for wood and $3 per month for supplies. (Courtesy of the author.)

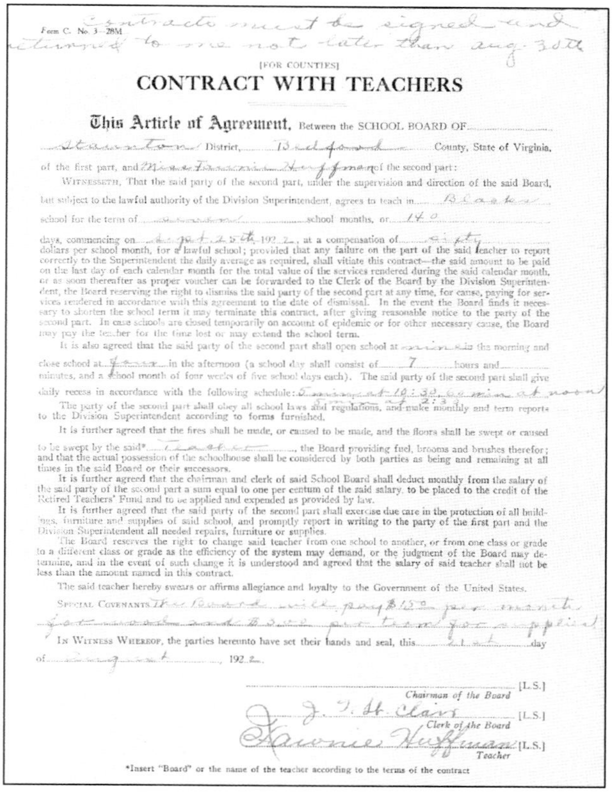

Form C. No. 3—28M

Contracts must be signed and returned to me not later than aug. 30th

[FOR COUNTIES]

CONTRACT WITH TEACHERS

This Article of Agreement, Between the SCHOOL BOARD OF Staunton District, Bedford County, State of Virginia, of the first part, and Miss Fawnie Huffman of the second part:

WITNESSETH, That the said party of the second part, under the supervision and direction of the said Board, but subject to the lawful authority of the Division Superintendent, agrees to teach in Blacks school for the term of .... school months, or 140 days, commencing on Sept. 25th 1922, at a compensation of sixty dollars per school month, for a lawful school; provided that any failure on the part of the said teacher to report correctly to the Superintendent the daily average as required, shall vitiate this contract—the said amount to be paid on the last day of each calendar month for the total value of the services rendered during the said calendar month, or as soon thereafter as proper voucher can be forwarded to the Clerk of the Board by the Division Superintendent, the Board reserving the right to dismiss the said party of the second part at any time, for cause, paying for services rendered in accordance with this agreement to the date of dismissal. In the event the Board finds it necessary to shorten the school term it may terminate this contract, after giving reasonable notice to the party of the second part. In case schools are closed temporarily on account of epidemic or for other necessary cause, the Board may pay the teacher for the time lost or may extend the school term.

It is also agreed that the said party of the second part shall open school at nine in the morning and close school at four in the afternoon (a school day shall consist of 7 hours and .... minutes, and a school month of four weeks of five school days each). The said party of the second part shall give daily recess in accordance with the following schedule: 5 min at 10:30, 60 min at noon, 5 min at 2:30

The party of the second part shall obey all school laws and regulations, and make monthly and term reports to the Division Superintendent according to forms furnished.

It is further agreed that the fires shall be made, or caused to be made, and the floors shall be swept or caused to be swept by the said* Teacher, the Board providing fuel, brooms and brushes therefor; and that the actual possession of the schoolhouse shall be considered by both parties as being and remaining at all times in the said Board or their successors.

It is further agreed that the chairman and clerk of said School Board shall deduct monthly from the salary of the said party of the second part a sum equal to one per centum of the said salary, to be placed to the credit of the Retired Teachers' Fund and to be applied and expended as provided by law.

It is further agreed that the said party of the second part shall exercise due care in the protection of all buildings, furniture and supplies of said school, and promptly report in writing to the party of the first part and the Division Superintendent all needed repairs, furniture or supplies.

The Board reserves the right to change said teacher from one school to another, or from one class or grade to a different class or grade as the efficiency of the system may demand, or the judgment of the Board may determine, and in the event of such change it is understood and agreed that the salary of said teacher shall not be less than the amount named in this contract.

The said teacher hereby swears or affirms allegiance and loyalty to the Government of the United States.

SPECIAL COVENANTS The Board will pay $1.50 per month for wood and $3.00 per term for supplies

IN WITNESS WHEREOF, the parties hereunto have set their hands and seal, this 21st day of August, 1922.

.... [L.S.]
Chairman of the Board

J. T. St. Clair [L.S.]
Clerk of the Board

Fawnie Huffman [L.S.]
Teacher

*Insert "Board" or the name of the teacher according to the terms of the contract

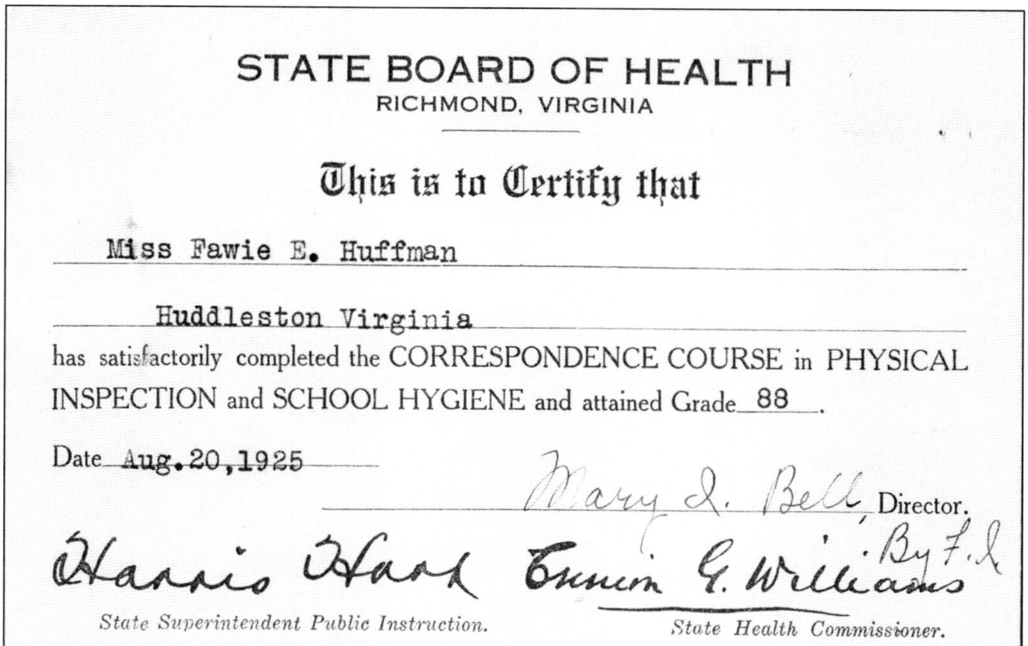

STATE BOARD OF HEALTH
RICHMOND, VIRGINIA

This is to Certify that

Miss Fawie E. Huffman

Huddleston Virginia

has satisfactorily completed the CORRESPONDENCE COURSE in PHYSICAL INSPECTION and SCHOOL HYGIENE and attained Grade 88.

Date Aug. 20, 1925

Mary I. Bell, Director.

Harris Hart — *State Superintendent Public Instruction.*

Ennion G. Williams By F. I. — *State Health Commissioner.*

Part of Huffman's necessary licensure was to complete a correspondence course in physical inspection and school hygiene. She passed with a grade of 88 and was officially certified on August 20, 1925. (Courtesy of the author.)

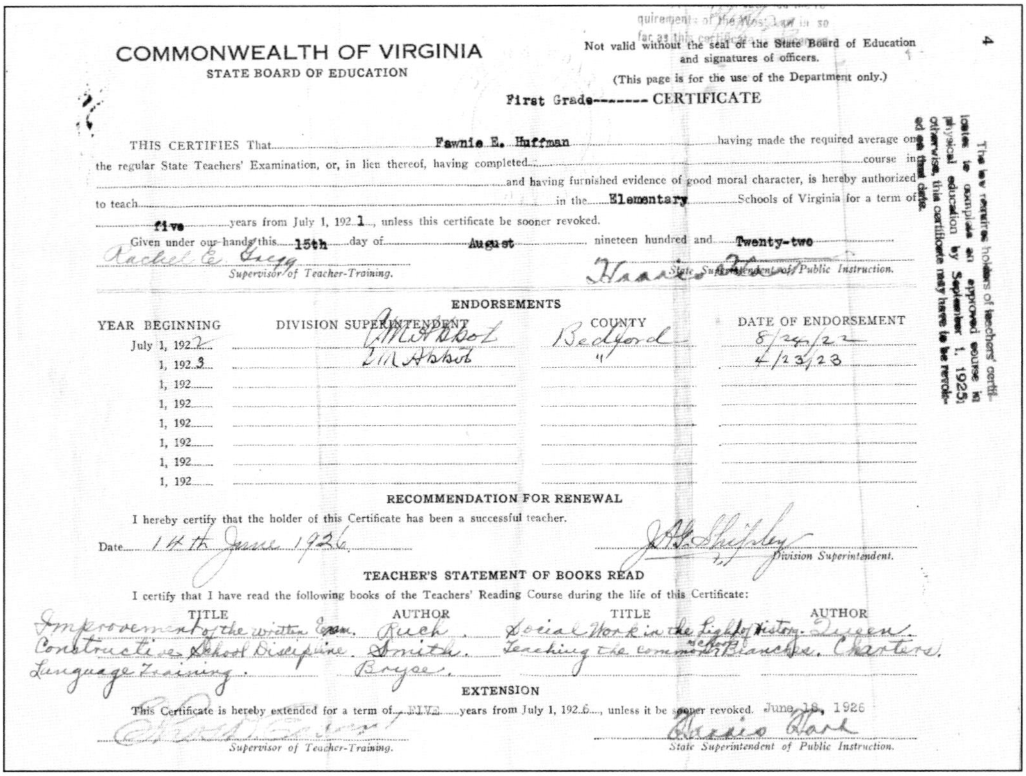

quirement of the ... Law in so far as this certificate ...

Not valid without the seal of the State Board of Education and signatures of officers.

(This page is for the use of the Department only.)

4

COMMONWEALTH OF VIRGINIA

STATE BOARD OF EDUCATION

First Grade------- CERTIFICATE

THIS CERTIFIES That Fawnie E. Huffman having made the required average on the regular State Teachers' Examination, or, in lieu thereof, having completed ... course in ... and having furnished evidence of good moral character, is hereby authorized to teach ... in the Elementary Schools of Virginia for a term of five years from July 1, 1921, unless this certificate be sooner revoked.

Given under our hands this 15th day of August nineteen hundred and Twenty-two

Rachel E. Gregg, Supervisor of Teacher-Training.

State Superintendent of Public Instruction.

The law requires holders of teachers' certificates to complete an approved course in physical education by September 1, 1925; otherwise, this certificate may have to be revoked on that date.

ENDORSEMENTS

YEAR BEGINNING	DIVISION SUPERINTENDENT	COUNTY	DATE OF ENDORSEMENT
July 1, 1922	J M Abbot	Bedford	8/24/22
1, 1923	J M Abbot	"	4/23/23
1, 192			
1, 192			
1, 192			
1, 192			
1, 192			
1, 192			

RECOMMENDATION FOR RENEWAL

I hereby certify that the holder of this Certificate has been a successful teacher.

Date 14th June 1926 — Division Superintendent.

TEACHER'S STATEMENT OF BOOKS READ

I certify that I have read the following books of the Teachers' Reading Course during the life of this Certificate:

TITLE	AUTHOR	TITLE	AUTHOR
Improvement of the written Exam.	Ruch.	Social Work in the Light of History.	Queen.
Constructive School Discipline.	Smith.	Teaching the common Branches.	Charters.
Language Training.	Bryce.		

EXTENSION

This Certificate is hereby extended for a term of FIVE years from July 1, 1926, unless it be sooner revoked. June 18, 1926

Supervisor of Teacher-Training.

State Superintendent of Public Instruction.

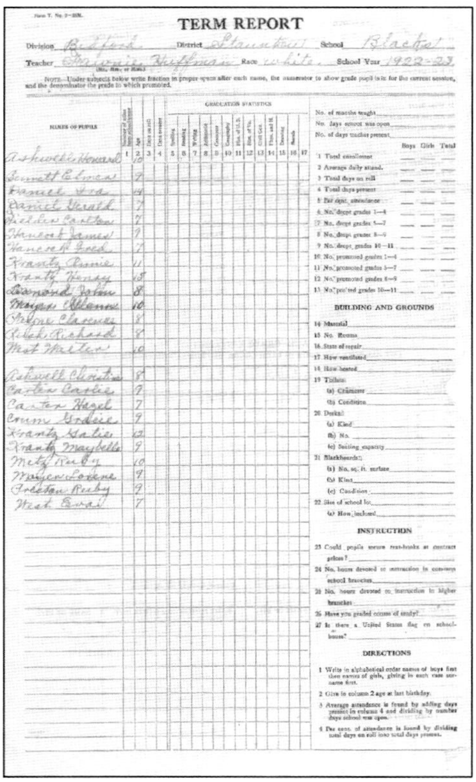

TERM REPORT

Division Bedford District Staunton School Blacks

Teacher Fawnie Huffman Race white School Year 1922-23

NOTE.—Under subjects below write fraction in proper space after each name, the numerator to show grade pupil is in for the current session, and the denominator the grade to which promoted.

NAMES OF PUPILS — GRADUATION STATISTICS

BUILDING AND GROUNDS

INSTRUCTION

DIRECTIONS

1. Write in alphabetical order names of boys first, then names of girls, giving in each case surname first.
2. Give in column 2 age at last birthday.

Huffman's State Board of Education Certificate as a first-grade teacher was issued for a five-year period, beginning July 1, 1921. In her spare time, she taught her father to read and write. (Courtesy of the author.)

Huffman's term report required information on each student, including number of miles "from the schoolhouse" and a space for grades in various subjects such as reading, writing, and arithmetic. Two facts are noteworthy: students were not only graded on specific subjects but also on "morals." Also the range of ages for her first-grade class was age 7 to 15. (Courtesy of the author.)

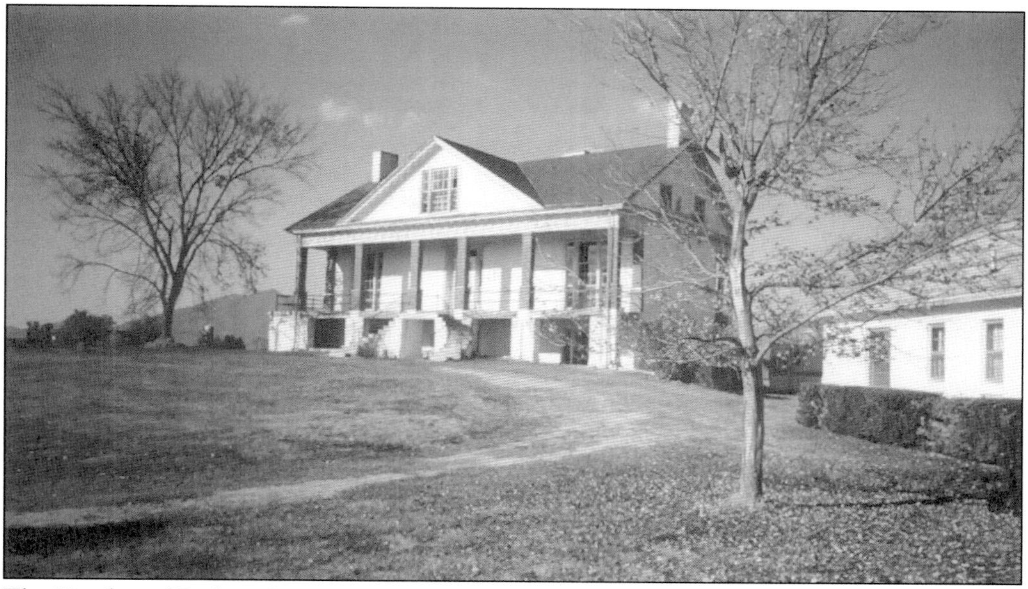

The Knights of Pythias Boys Home near Liberty High School became a private school called Otterburn Academy in 1967. After it moved to a new facility along State Route 122 on what is now Wells Road, it grew to a grade K–12 facility. The last graduating class received their diplomas in 1979. (Courtesy of the author.)

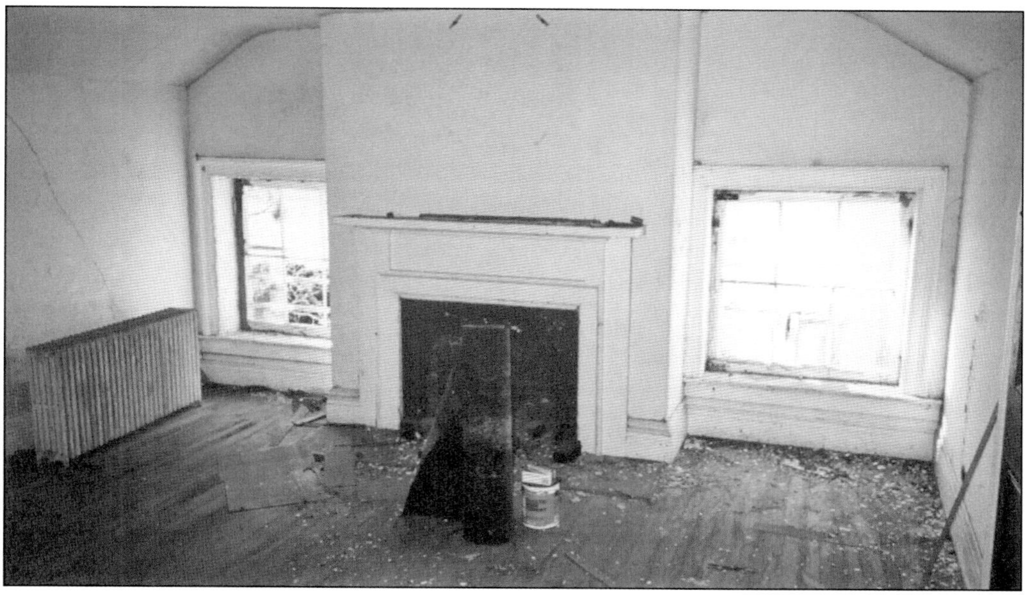

One of two Otterburn first-grade classes was held in this former upstairs bedroom in the former Knights of Pythias Boys Home in 1967. (Courtesy of the author.)

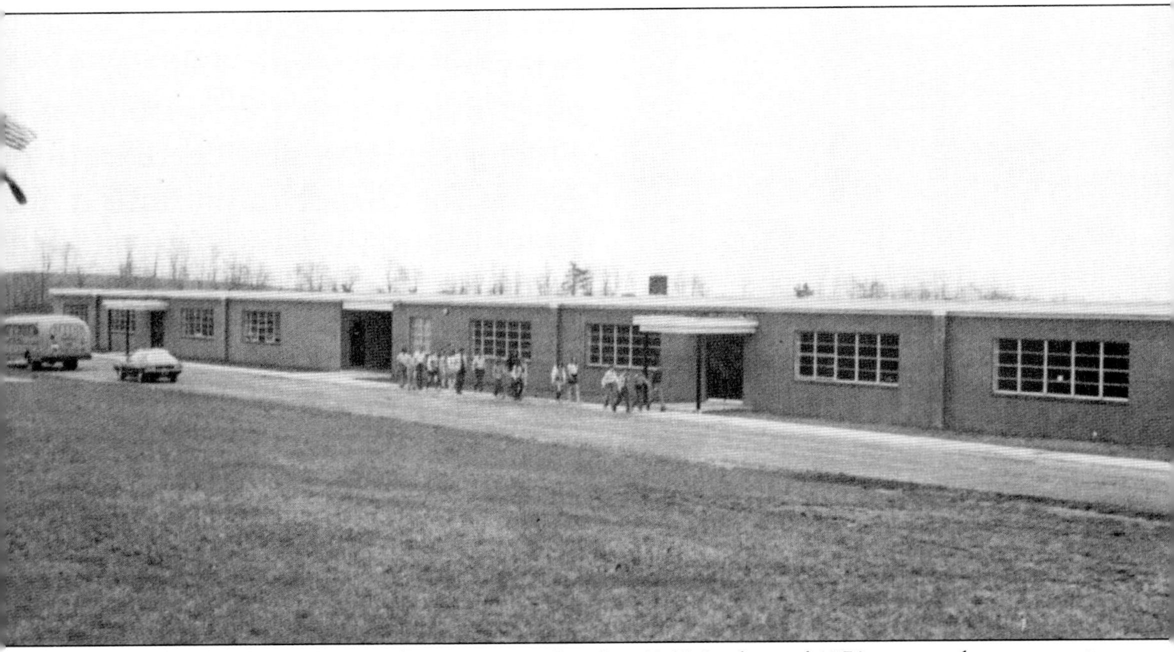

The second Otterburn Academy was completed in 1968. In the mid-1970s, a complete gymnasium was added. (Courtesy of the author.)

Seven

EARLY SMITH MOUNTAIN LAKE

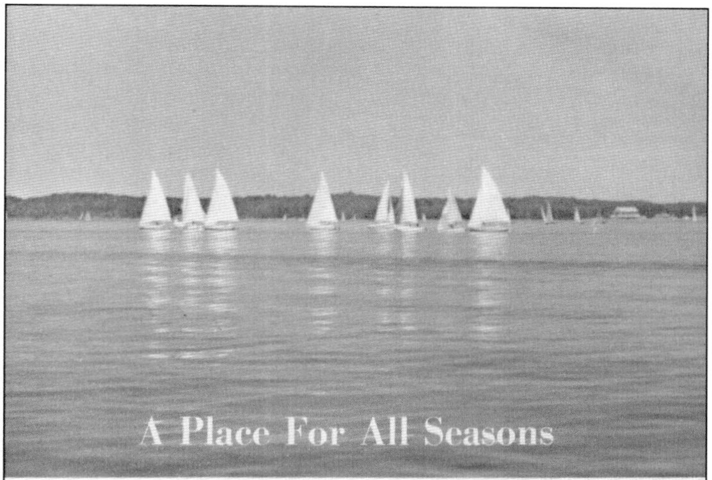

A Place For All Seasons

Aside from its rustic beauty and tranquil charm, Smith Mountain Lake's foremost attribute is its recreational aspect. Every outdoor sport is enjoyed here, and is, in fact, enhanced by this spectacular setting. Even those who are not sports minded will find a stay at this western Virginia Shangri-La a relaxing and refreshing interlude.

A number of restuarants and motels offer attractive accommodations amid the peaceful setting of the Blue Ridge foothills. Cedar Key Resort near the dam rents lodges by the day or week and guide service is also available here. This facility is open year around and caters especially to visiting sportsmen although its attractive decor offers an inviting respite to any vacationist. Limousine service is provided to airports at Lynchburg and Roanoke.

Approximately 20 campgrounds are scattered around the lake with a varying degree of facilities and accommodations.

The Commonwealth of Virginia is developing a state park on the north shore of the lake off state route 626. The 1,200 acre tract is located partly on land donated by Appalachian Power Company and partly on property purchased by the state. Tentative plans for construction include a large marina, cottages and a restuarant.

Some $265,000 has been approved by the Virginia Commission of Outdoor Recreation for construction of roads into the park. An additional $50,000 was appropriated for planning by the 1974 General Assembly.

Boats and motors are available at many of the marinas around the lake and the state maintains five concrete launching ramps for the sportsman's convenience. These are located off state route 626, south of the dam and at Brooks Mill Bridge in Franklin County, at Hardy's Ford Bridge and at the end of state route 823 in Bedford County. There is also a proposed ramp off state route 734 on the north shoreline near the dam.

For the boat enthusiast or water skiier Smith Mountain Lake offers an infinite variety of scenery and plenty of room on its 20,000 acres.

The lake winds some 40 miles through the picturesque mountain region and has more than 500 miles of shoreline composed of nooks, coves and points. In addition to the public ramps, most of the marinas have launching facilities, as well as food and marine fuel.

A power squadron comprised of power craft owners has been organized at the lake. For sail boat owners there is the Visa Yacht Club (Virginia Inland Sailing Association) which sponsors over 30 races each year plus two invitational regattas and operates a club

Even as early as 1974, the publication year for a booklet produced by the Smith Mountain Lake Association called *Smith Mountain Lake, Virginia*, the lake was "a place for all seasons." Not only were the counties surrounding the lake hosting a growing number of seasonal tourists each year, but many people were also making the area their permanent home. Note that this page says, "Even those who are not sports minded will find a stay at the western Virginia Shangri-La a relaxing and refreshing interlude." (Courtesy of Lib and Charlie Walker.)

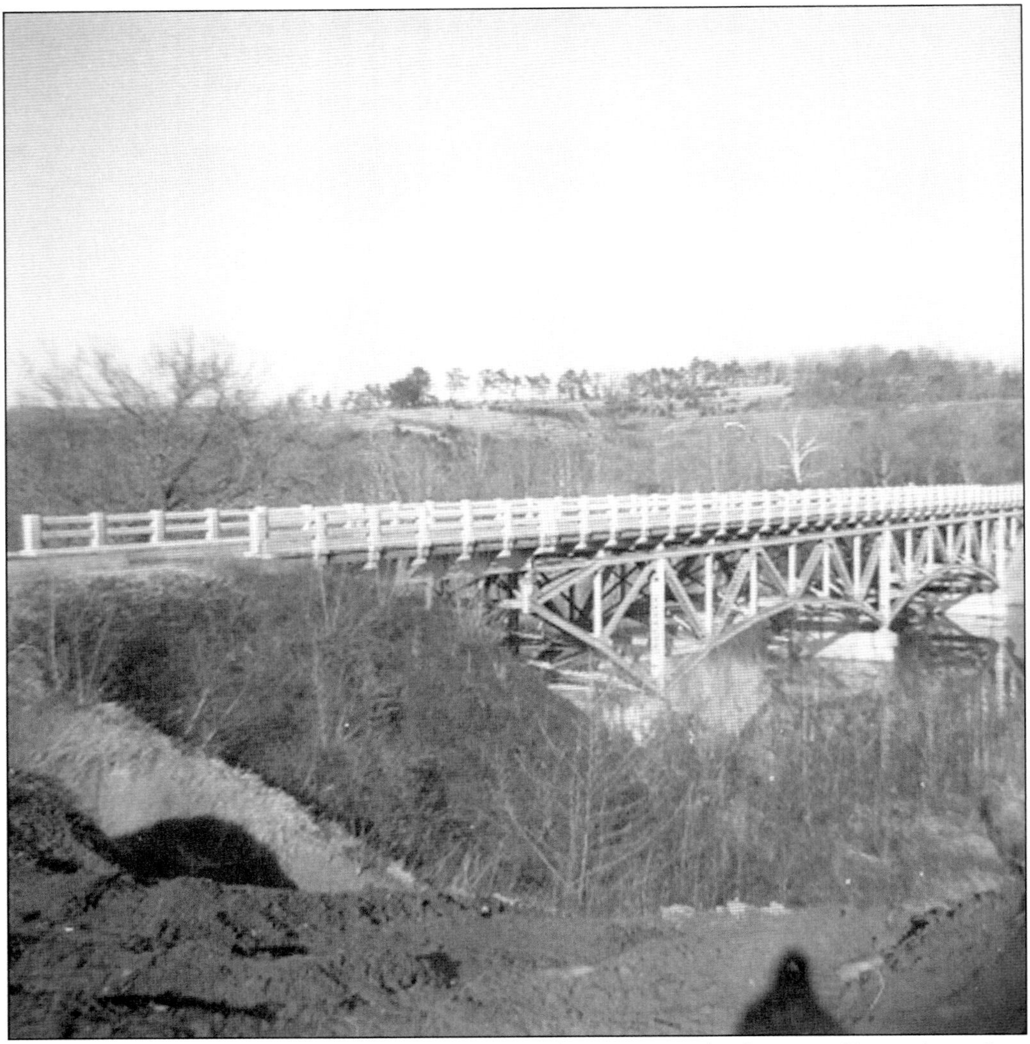

This undated photograph shows the Hales Ford Bridge spanning the Staunton River. According to a February 6, 1964, *Bedford Bulletin-Democrat* newspaper article, "Last week's announcement that the Smith Mountain Lake was filling steadily, with backed up waters five and a half miles beyond Hale's Ford Bridge on Route 122, attracted several hundred visitors to the bridge and its vicinity in Bedford County Sunday." Many people stood on the older bridge to gaze up at the newer one and watch the waters rise. Appalachian Power built six bridges and 22 miles of road at a cost of $2.3 million for its Smith Mountain Lake project. (Courtesy of Shirley Ring.)

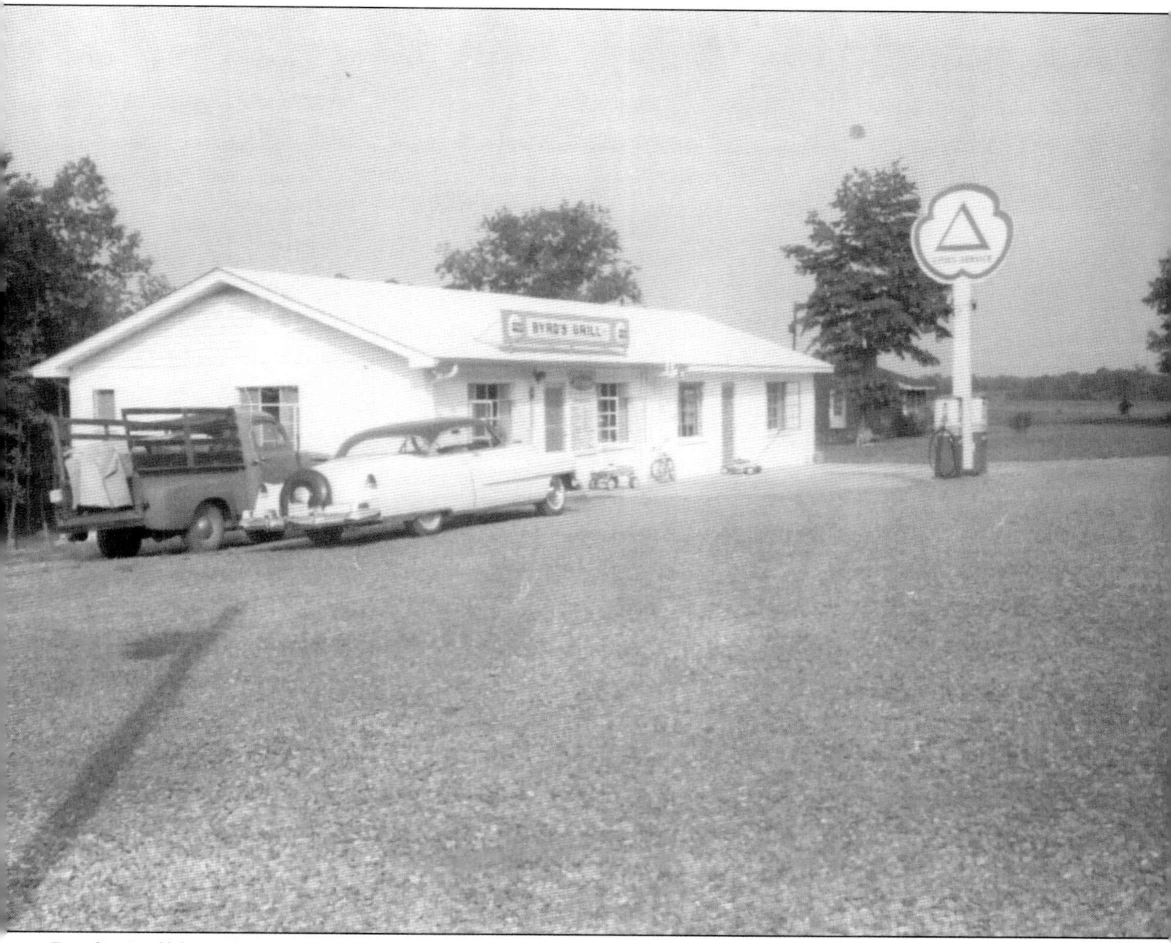

Byrd's Grill later became Byrd's Restaurant and was affectionately known as "the Byrd Cage." It was one of only a few restaurants in Bedford County in the 1950s within a few miles of the lake. Byrd's, located on the present-day Diamond Hill Road, served both tourists and travelers with typical, home-cooked Southern fare. Today the building serves the community as the site of the Smith Mountain Eagles Lodge. (Courtesy of Clifton, Nancy, David, and Steve Byrd.)

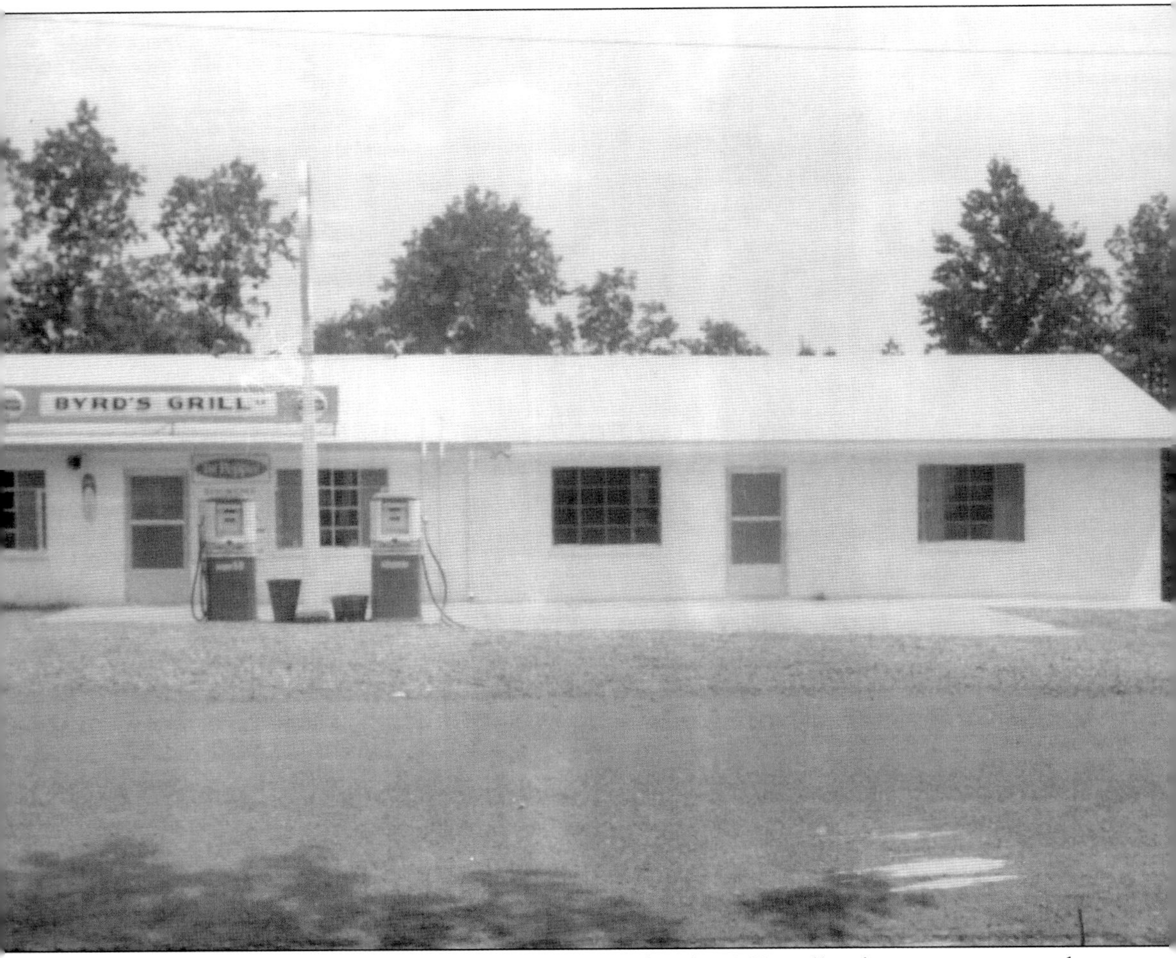

Byrd's Grill, as it was called in the late 1950s and early 1960s, offered more to patrons than a cheeseburger and an order of fries; it also featured a huge dining room with soft-leather booths, a jukebox, and a game room featuring a full-size pool table. (Courtesy of Clifton, Nancy, David, and Steve Byrd.)

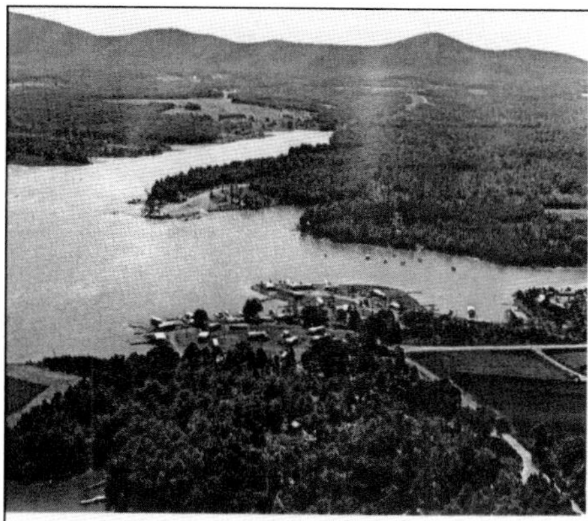

SMITH MOUNTAIN LAKE DIRECTORY

MARINAS

SMITH MOUNTAIN DOCK 1
Owner: Fred Munson
Rt. 626
Penhook, Va. 24137
Phone: (804) 927-5100
Lodging, fishing licenses, boat rental, mooring, scuba tanks filled and rented, ski kite sales.

LAKESIDE MARINA 2
Owners: Richard and Barbara Dill
Rt. 626
Mail: Rt. 1, Box 103
Penhook, Va. 24137
Phone: (804) 927-5380
Floating and stationary docks, boat houses, electric hoists and boat salvage. Also boat moorings of any type, boat rental, fishing and camping supplies, snack bar and groceries. Campground facilities are available.

PENHOOK BOAT DOCK 3
Owner: Ron McLaughlin
Rt. 660
Mail: Rt. 1
Penhook, Va. 24137
Phone: (804) 576-2737
Launching, boat moorings, rentals, fishing supplies, live bait.

FOXPORT XIV MARINE CAMP 5
Co-Owner: Wayne Blount
9 miles from Rt. 122 at Burnt Chimney via Rt. 670 and 1320
Mail: Rt. 1, Box 136-B
Wirtz, Va. 24184
Phone: 721-2451 or 721-9991
Marina, lodging, launching, groceries, tackle, bait, marine supplies, boat rentals, mooring rentals, campsites, beach area.

CAPE CARLYN MARINA 7
Owner: Carl Higginbotham
Rt. 1, Box 153
Goodview, Va. 24095
Phone: (703) 297-4919
Boat and motor sales and services, accessories, launching ramp, crane service, slip rental, boat storage, boat and motor rentals, gas and oil, fishing supplies, trailer spaces, cottages and apartments for rent.

BAY ROC MARINA AND DRY STORAGE 8
Owner: Howard L. Propst
Rt. 634 at Hardy Ford Bridge
Mail: Rt. 1, Box 91
Hardy, Va. 24101
Phone: (703) 890-2194
Boat and motor sales and service, boat storage-both wet and dry, fishing and water sport supplies. We make service calls on the lake.

SMITH MOUNTAIN YACHT CLUB 12
Owner: Richard L. Arnold
P. O. Box 495
Moneta, Va. 24121
Phone: 297-4484
Complete Marina facilities, boat sales and service, docking, gas, groceries, ice, fishing, fishing tackle, live bait, marine supplies, free launching ramp, large shop, complete repair and overhaul service, 24,000 lb. Travel life, radio communications—we monitor Channel 9—citizens band, open all day—every day, year round, clean, modern restrooms and showers. Everyone welcome.

GROSS POINT MARINA 16
Owner: H. A. Gross
Rt. 1, Huddleston, Va. 24104
Phone: 297-9811
Rooms for rent, groceries, picnic supplies, restaurant, picnic tables, launching ramp, fishing and boat rentals, fishing pier, anchorage, slip rental, fish bait and supplies, gas and oil for boats and cars.

SAUNDERS MARINA 17
Owner: Jimmie R. Saunders
Rt. 1, Huddleston, Va. 24104
Phone: 297-4412
Boat and motor sales and service, docking facilities, crane service, slip rental, moorage, launching ramp, boat storage, fishing, sport and family boats for rent, fishing equipment and supplies, fishing license, lake guides, picnic and camping supplies, groceries, cabins for rent.

CAMPGROUNDS

PELICAN POINT CAMPGROUND 4
Owners: Ralph and Ginny Zahn
Rt. 957
Union Hill, Va. 24176
Phone: (703) 576-2019
Season from April 12 through November 3. Shaded campsites with over 6,000 feet of shore line. Write or call for free brochure.

CRAZY HORSE CAMPGROUNDS 6
Owners: Karl and Carol Kuhn
Rt. 616, 6 miles east of Rt. 122
Mail: Rt. 1, Box 178 A
Moneta, Va. 24121
Phone: (703) 721-2792
Open year round for fishing, boating, hunting and camping. Tent areas, playground, miniature golf, recreation hall, picnic areas, boat rentals, dock and launching facilities, LP gas, ice and swimming pool.

WATERFRONT PARK 10
Owner: Ted R. Leneski
Rt. 1, Moneta, Va. 24121
Phone: 297-5364
Wooded trailer and tent sites, water and electricity, two modern bath houses, dumping station, grocery store, launching ramp, swimming, nature trails, gasoline station.

SAFARI CAMPGROUND 14
Call TOLL FREE for reservations and opening date 800-525-8585, local calls 297-5433.
We have 115 acres, a mile of waterfront, 60 sites, full hookups, shaded sites, sculptured swimming pool, teen lounge, children's playground, food and equipment store, ice, boat launch ramp, boat and canoe, bicylce, rental, miniature golf, coin laundry, immaculate wash rooms with hot showers and dressing rooms. Brochure on request: P. O. Drawer A, Huddleston, Va. 24104.

EAGLE'S ROOST 15
Owner: D. R. Cosgrove
Rt. 1, Huddleston, Va. 24104
Phone: 297-3181
Wooded camp sites, hot and cold showers, laundrymat, groceries, ice, launching ramp, family and fishing boat rentals, bait.

REAL ESTATE

LAKESIDE REALTY 9
Waterfront homes and homesites.
Melvin Johnston, Broker
Phone: 297-4957 or 721-2355

MOTELS and RESTAURANTS

LAKE SHORE RESTAURANT 11
Owner: Mrs. J. C. Peters
Operator: J. C. Peters
Rt. 122, 1 mile East of Hale's Ford Bridge
Phone: 297-9881
Breakfast, Lunch, Dinner served. Fresh seafood every Wednesday nite in season. ABC —On and Off.

AIRPORT

SMITH MOUNTAIN AIRPORT, INC. 13
Moneta, Va. 24121
Phone: (703) 297-5181
3,050 ft. paved runway, 100 octane aviation gasoline, tie-downs, aircraft rentals, flight instruction, maintenance (major). Operational 24 hours daily. Cottage rentals.

FISHING RESORT

CEDAR KEY FISHING RESORT 18
Owners: Hester and Leo Bourassa
Rt. 608
Huddleston, Va. 24104
Phone: (703) 297-5118
Cabins completely equipped for housekeeping, boat rentals and clubhouse.

19 SMITH MOUNTAIN DAM – INFORMATION CENTER – RECREATION AREA – PICNICKING, FIREPLACES, REST ROOMS
20 PROPOSED STATE PARK AREA

By the early 1970s, growth along the shoreline was beginning to boom. Included in this chapter are several advertisements from some of the new businesses that were springing up along the 500-mile shoreline of the lake in *Smith Mountain Lake, Virginia.* The booklet's statistical abstract lists "4 establishments with overnight facilities on the lake; unlimited facilities within commuting distance." Already the Smith Mountain Lake Airport had been constructed near Moneta with a 3,000-foot runway. Notable organizations included the previously mentioned Lake Association, VISA Yacht Club, Coast Guard Auxiliary, REACT, Smith Mountain Lake Ministry, 4-H Club, and the Reservoirs Regional Planning and Economic Development Committee. While this lake directory is on one printed page in 1974, a glance through the enormous Web site www.visitsmithmountainlake.com shows just how much recreation, commerce, and residential living has grown in the past 35 years. (Courtesy of Lib and Charlie Walker.)

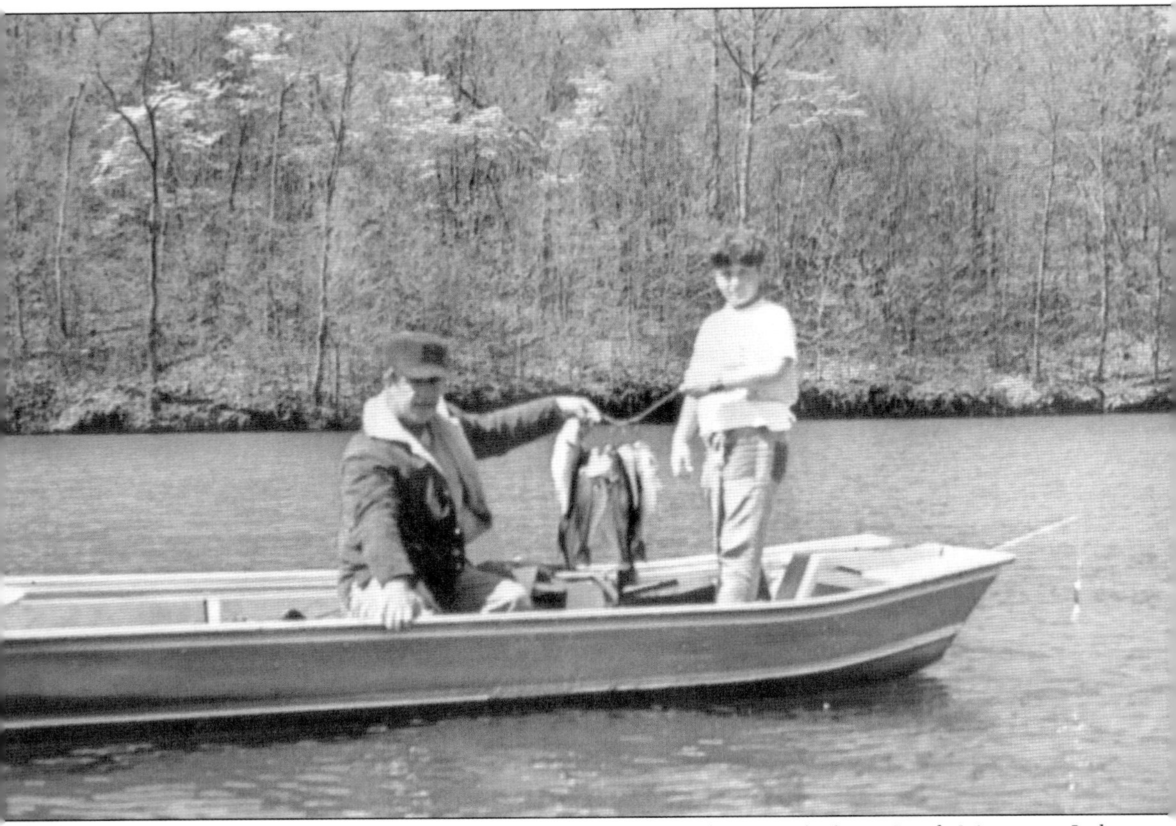

A father and son are enjoying a day of fishing in a photograph from *Smith Mountain Lake, Virginia*. The first few generations of residents who moved to Bedford County to live on or nearby Smith Mountain Lake became part of the farming community. For the author, tourists, boats, and water skis were as much a part of the experience of farm life during this era as were tractors, barns, and milk cows. In a style reminiscent of general store social interaction, it was not uncommon, for example, to see several farmers gathered outside a marina's storefront, soft drinks in hand, conversing with one another as well as with the tourists who were there to buy tackle and other necessities for a day's excursion on the lake. Actually, many early-20th-century general stores in Bedford County were still in operation in the 1960s and 1970s and profited from the increased tourist traffic. (Courtesy of Lib and Charlie Walker.)

One of the first advertisements for Pelican Point Campground is shown in *Smith Mountain Lake, Virginia*. The campground began operations in 1974. (Courtesy of Lib and Charlie Walker.)

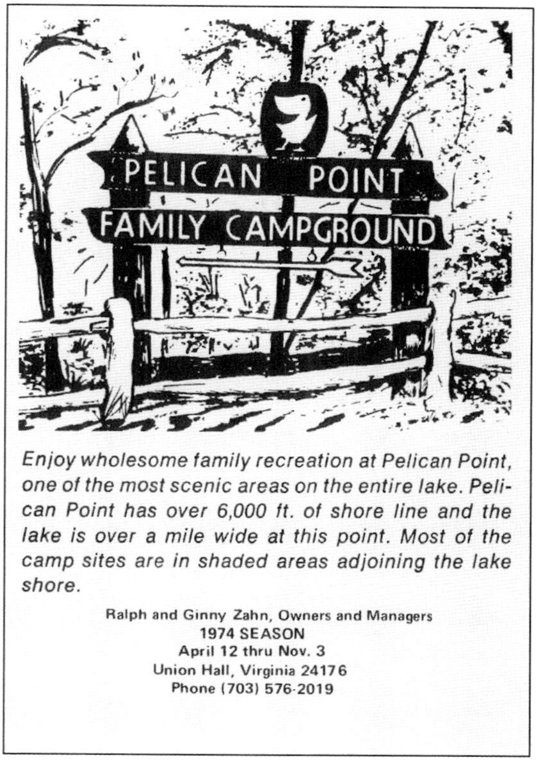

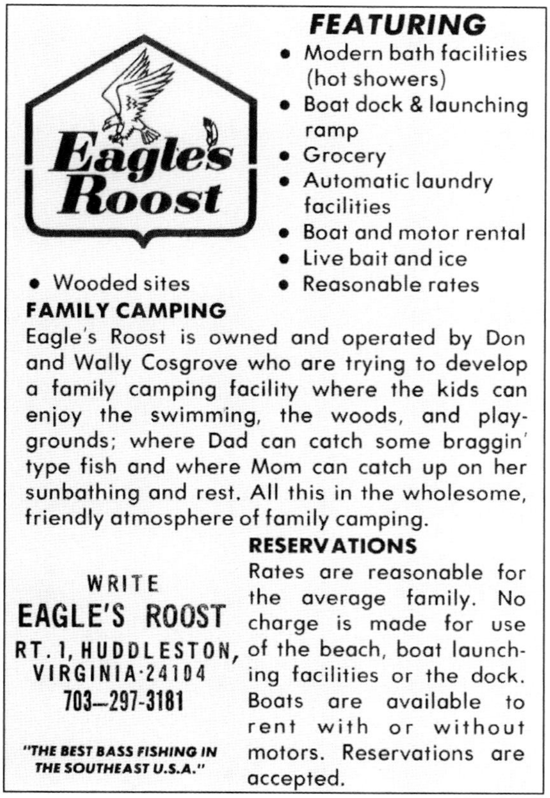

Note that this advertisement for Eagle's Roost Campground, which is another older lake business still in operation, boasts "wooded sites," a necessary attraction for tourists wanting to enjoy the great outdoors. (Courtesy of Lib and Charlie Walker.)

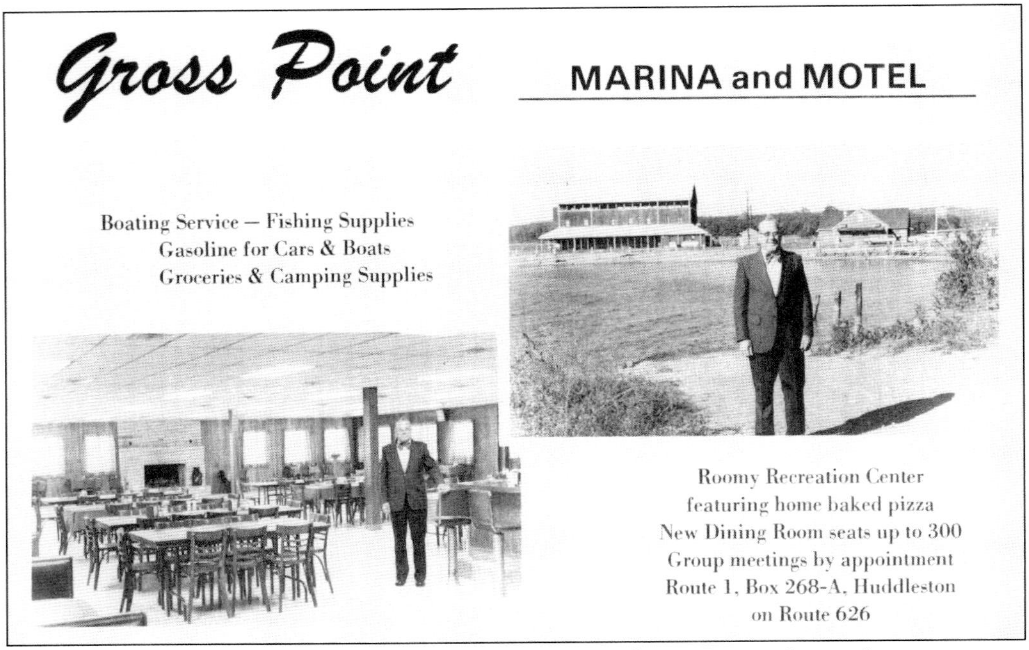

Gross Point Marina, along Smith Mountain Lake Parkway Drive, featured a restaurant, motel, "gasoline for cars & boats, groceries and camping supplies." (Courtesy of Lib and Charlie Walker.)

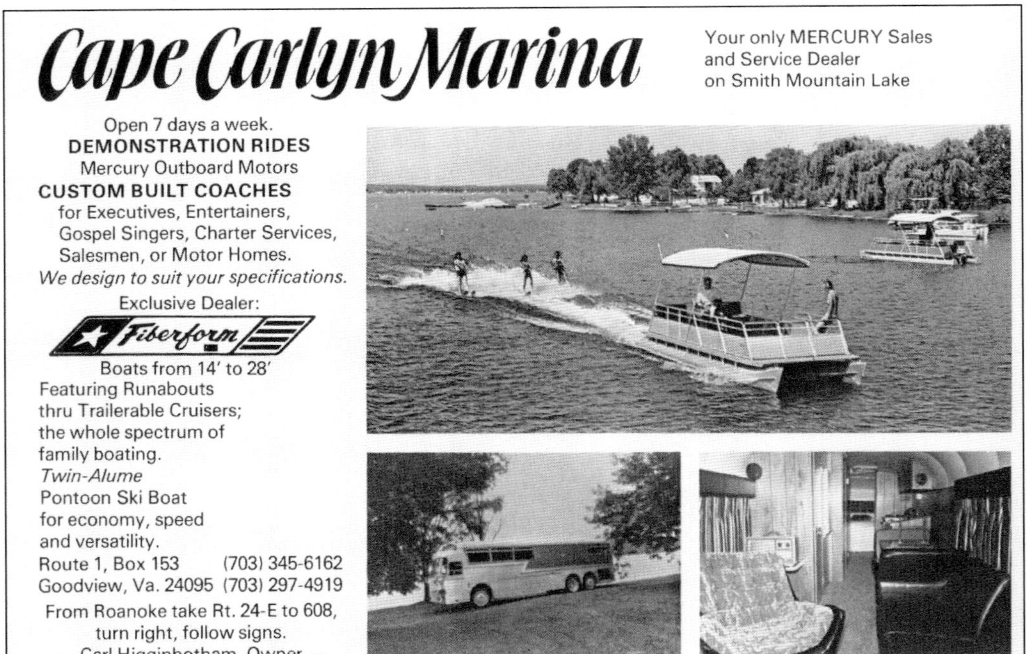

Cape Carlyn Marina was one of the first marinas on Smith Mountain Lake. According to www.lynhigginbotham.com, a Long and Foster realtor Web site, the venture began in 1963 and was the first for Carl and Lyn Higginbotham, thus the business name, Carl–Lyn. As the advertisement says, they were, at the time, the only sales and service dealer for the popular Mercury boat brand. (Courtesy of Lib and Charlie Walker.)

Eight

COUNTY SEAT

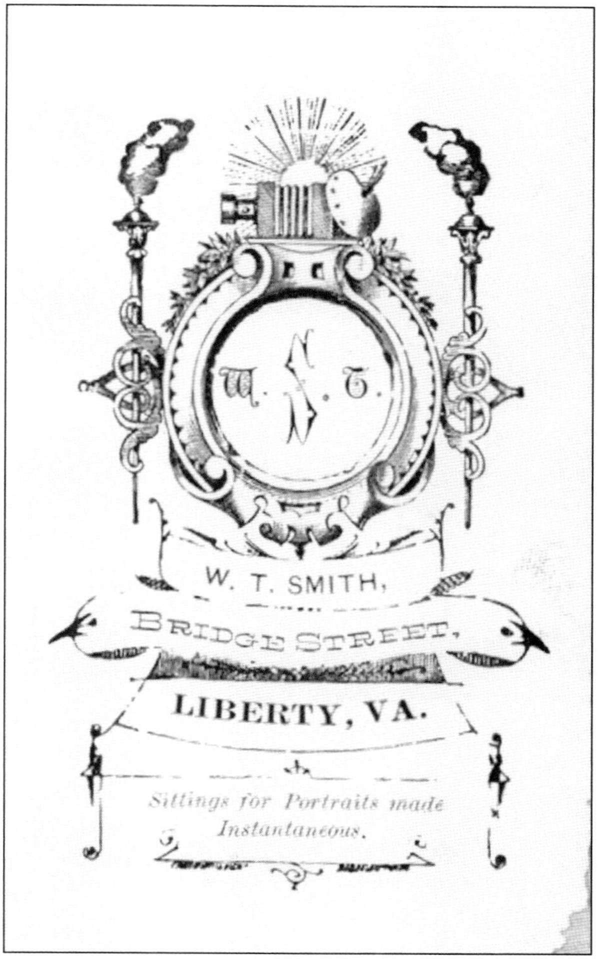

The town of Liberty, Virginia, was established in 1782. The town of Liberty became the town of Bedford in 1870. Bridge Street photographer W. T. Smith used this postcard to advertise his business with the slogan, "Sittings for Portraits made Instantaneous." (Courtesy of Dr. James Rucker Jr.)

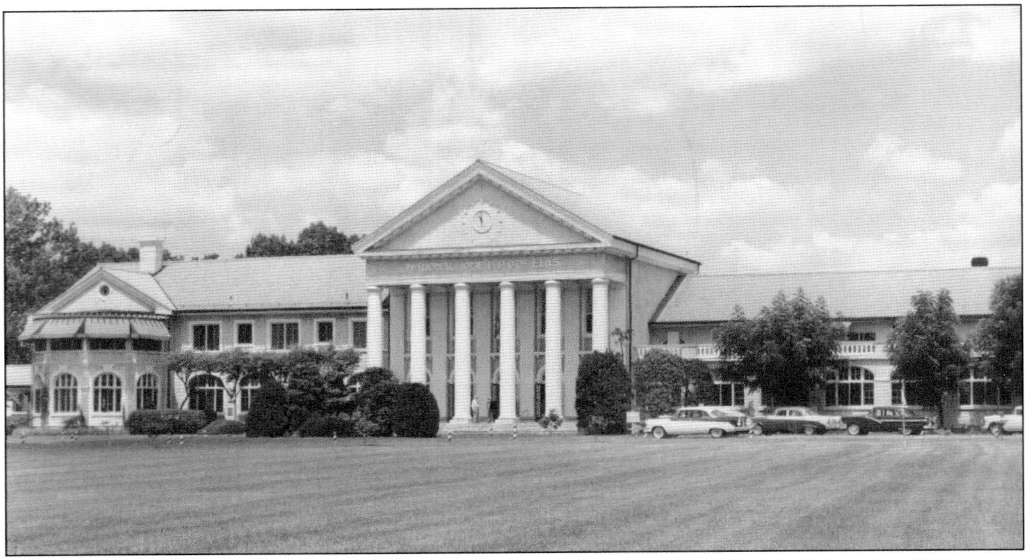

The Elks National Home in Bedford was first housed in the Hotel Bedford, but in 1916, a new building was erected on the current site at a cost of $450,000. According to the *Bedford Bulletin*, "The new home was dedicated July 8, 1916 by future President of the United States, Warren G. Harding, a Past Exalted Ruler of the Cleveland, Ohio Lodge, who at that time, was a U.S. Senator." (Courtesy of Frank Ewald.)

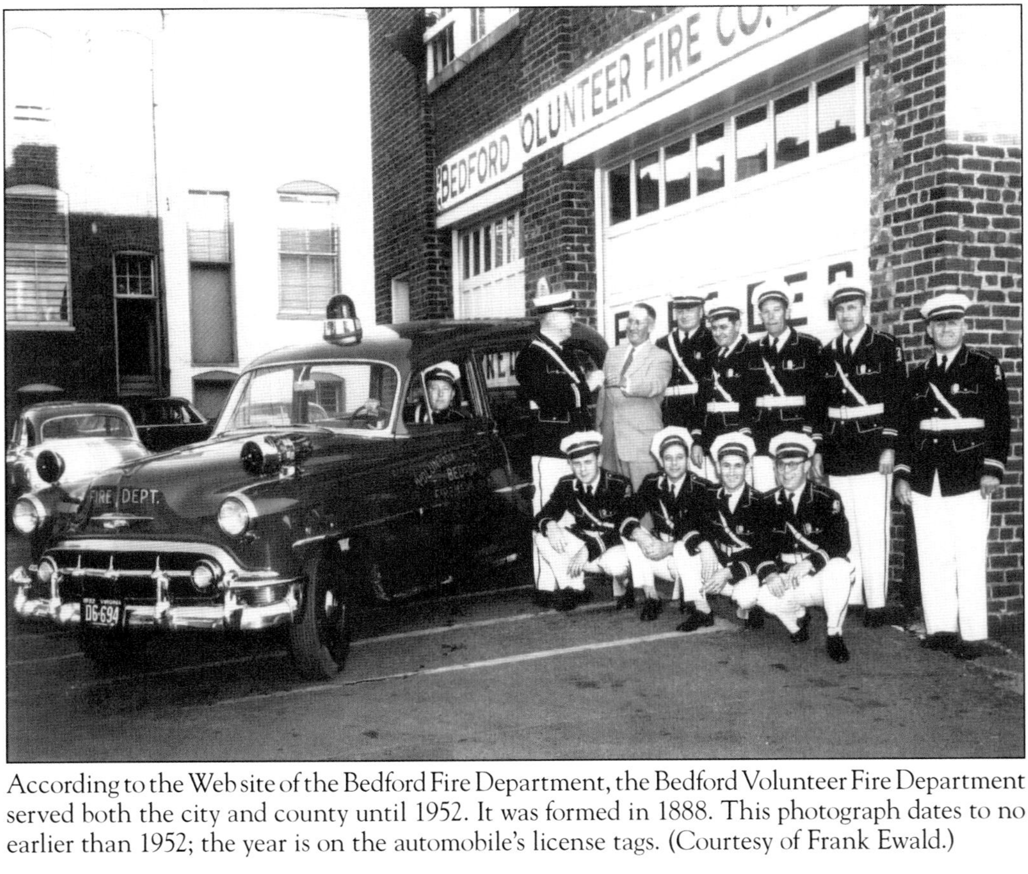

According to the Web site of the Bedford Fire Department, the Bedford Volunteer Fire Department served both the city and county until 1952. It was formed in 1888. This photograph dates to no earlier than 1952; the year is on the automobile's license tags. (Courtesy of Frank Ewald.)

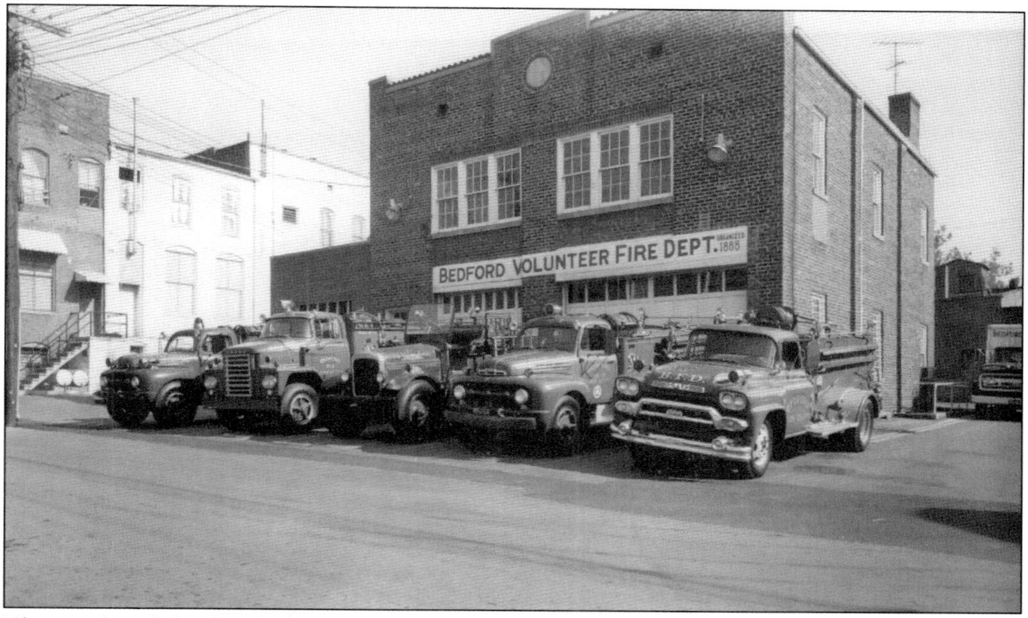

The trucks of the Bedford Fire Department are lined up outside the firehouse in this 1950s photograph. (Courtesy of Frank Ewald.)

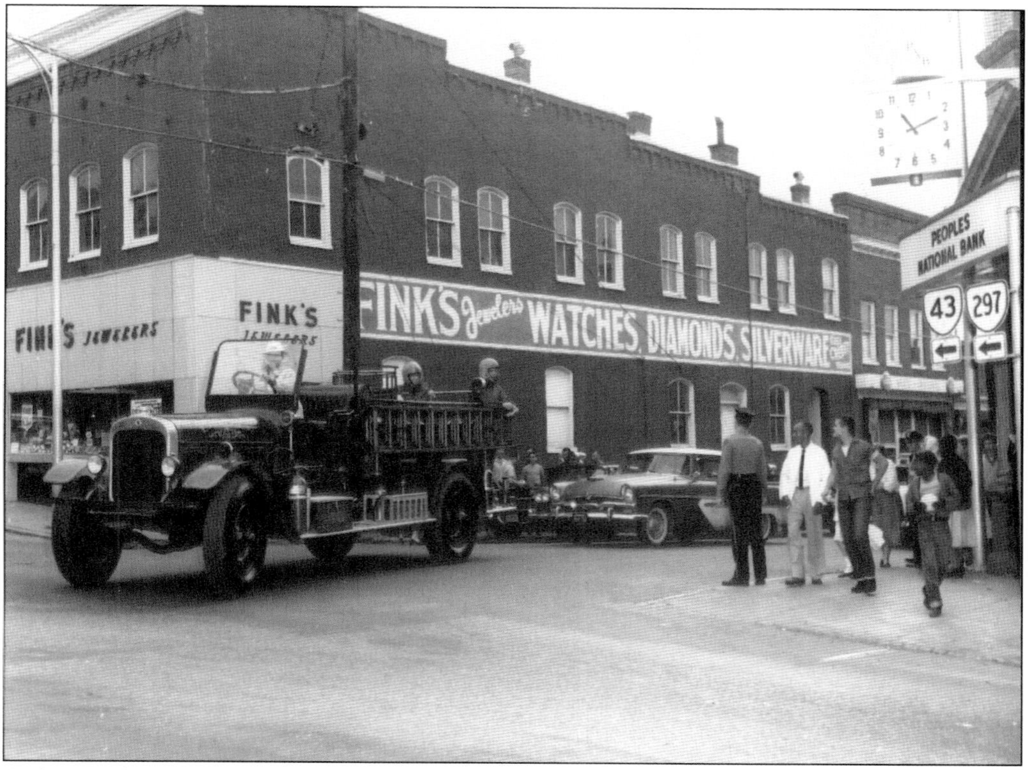

A fire truck crosses Main Street in Bedford in this 1950s photograph and is heading north. Note that Fink's Jewelry, a Roanoke-based firm, once had a store on the corner at 100 East Main Street. Fink's opened here in 1949. The renovations to the earlier structure as seen in this photograph occurred in 1954. (Courtesy of Frank Ewald.)

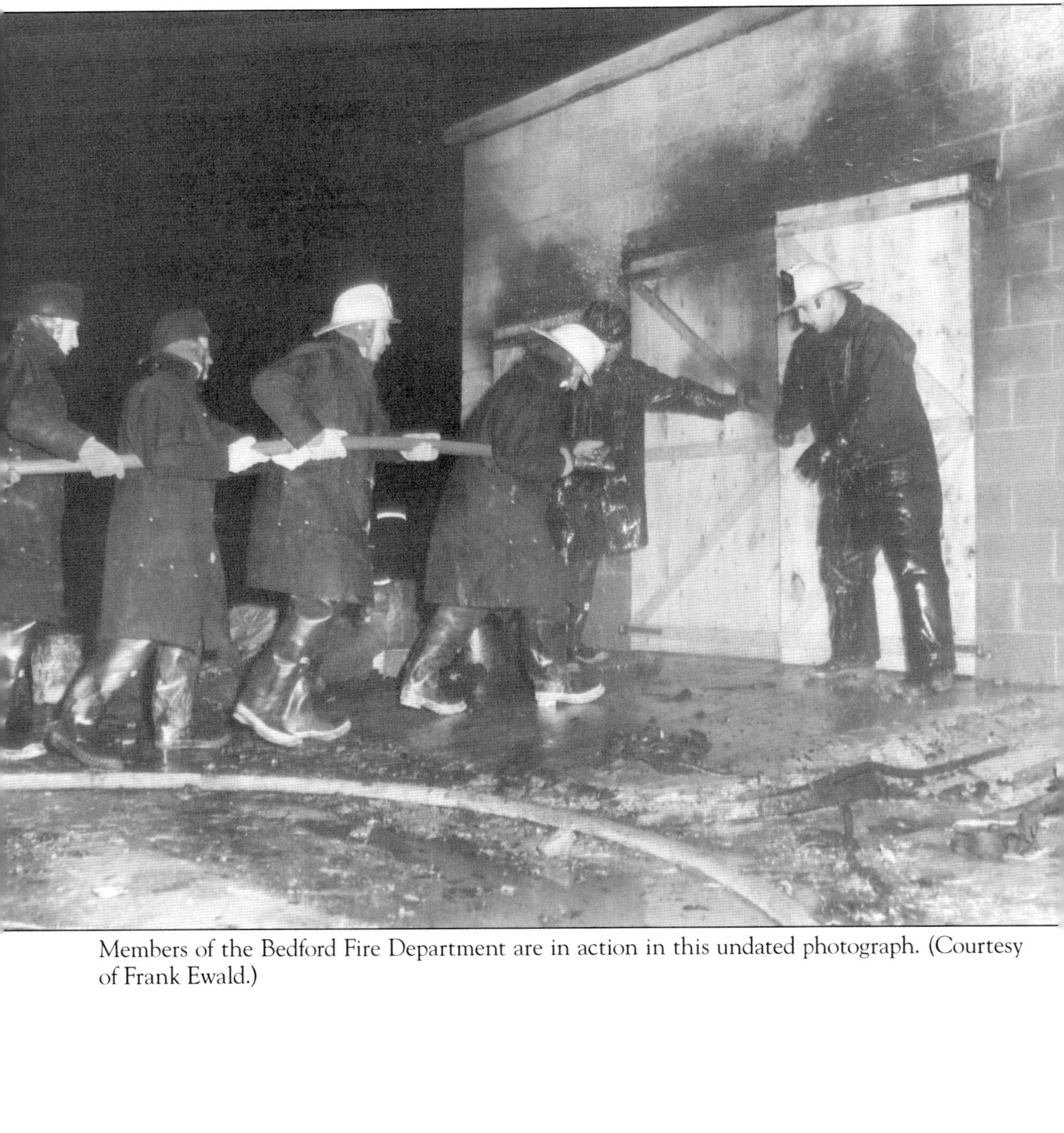

Members of the Bedford Fire Department are in action in this undated photograph. (Courtesy of Frank Ewald.)

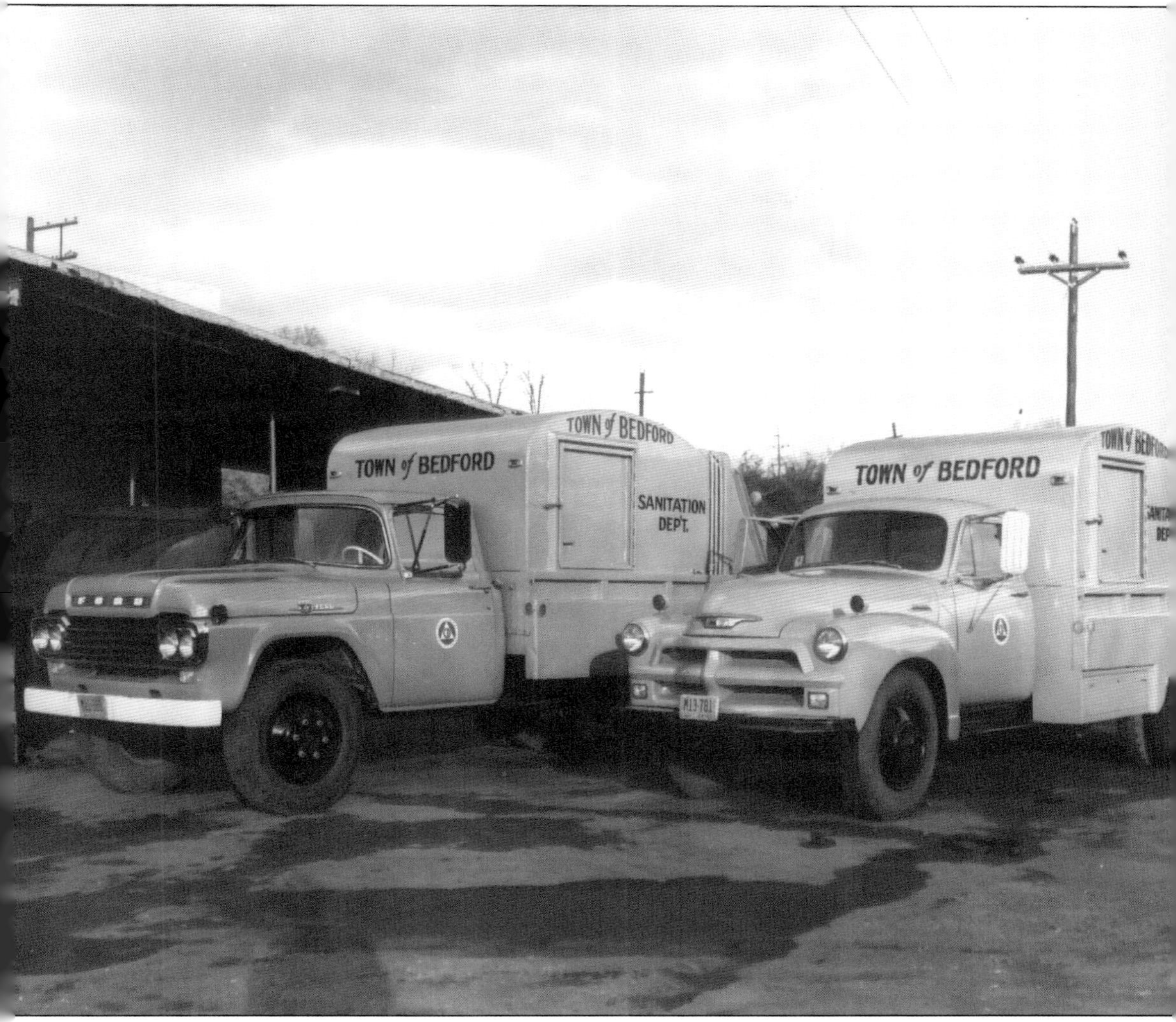

These two trucks show how even trash pick-up has changed since the 1950s. There were no mechanical arms on these Town of Bedford Sanitation Department trucks to pick up heavy green metal containers, only the human type to hoist a much smaller variety of trash can. (Courtesy of Frank Ewald.)

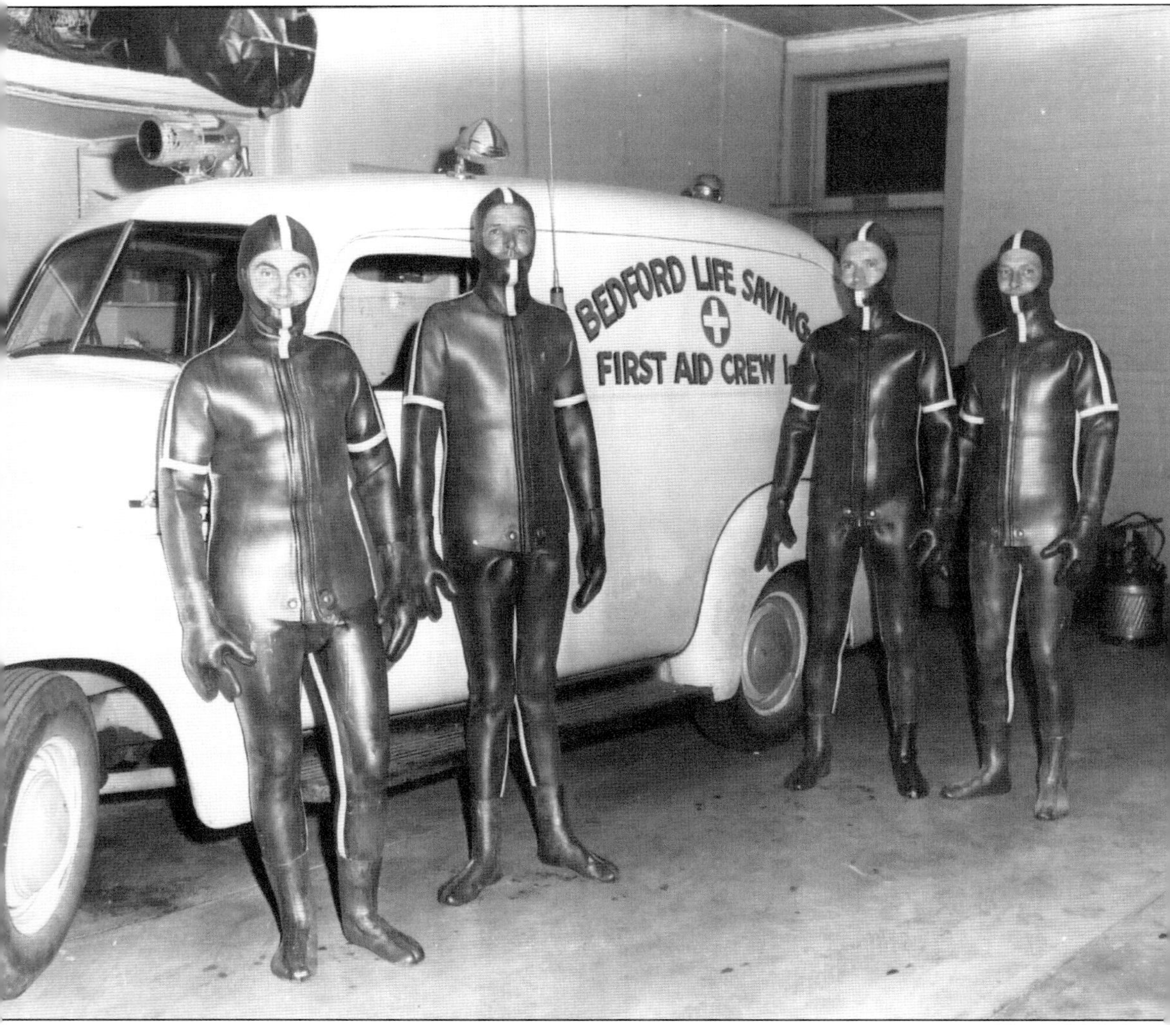

No, these are not aliens from outer space and this is not a scene from a 1950s science fiction film. Instead, these men are members of the Bedford Life Saving and First Aid Crew posing in outfits designed for underwater rescue. According to the *Bedford Bulletin*, this unit began in 1943 with 15 members and one Ford truck. (Courtesy of Frank Ewald.)

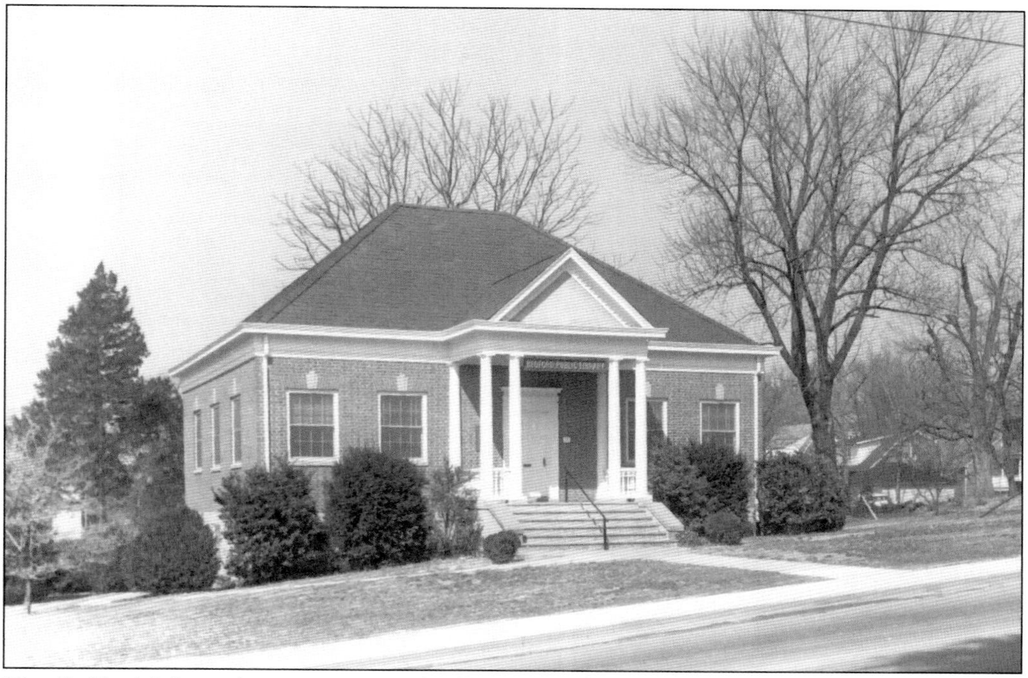

The Bedford Library has its roots in the Women's Library Association of Bedford, according to the library Web site. On November 30, 1900, this association began a library in a two-room office at the site of this structure on Bridge Street. Built in 1941, this facility was first open to the public in 1942. (Courtesy of Frank Ewald.)

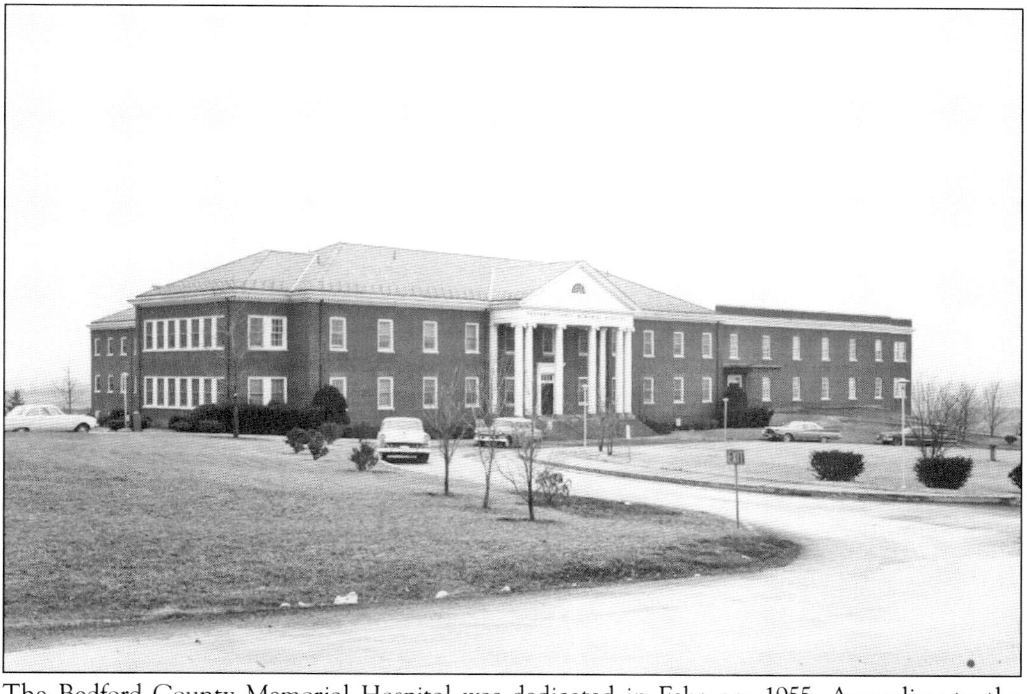

The Bedford County Memorial Hospital was dedicated in February 1955. According to the *Bedford Bulletin*, Dr. Henry B. Mulholland, who was the guest speaker, said, "The baby has been delivered, now you must see that it is nourished and grows." (Courtesy of Frank Ewald.)

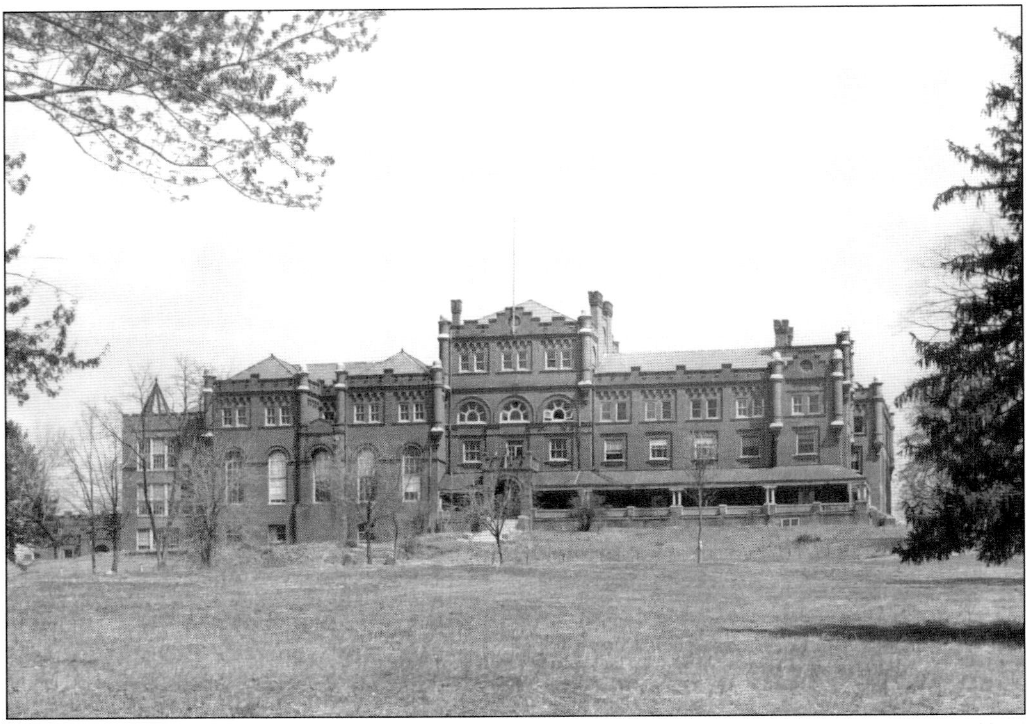

An early Town and County of Bedford pamphlet says, "Liberty Academy is one of Bedford County's Consolidated Schools." (Courtesy of Frank Ewald.)

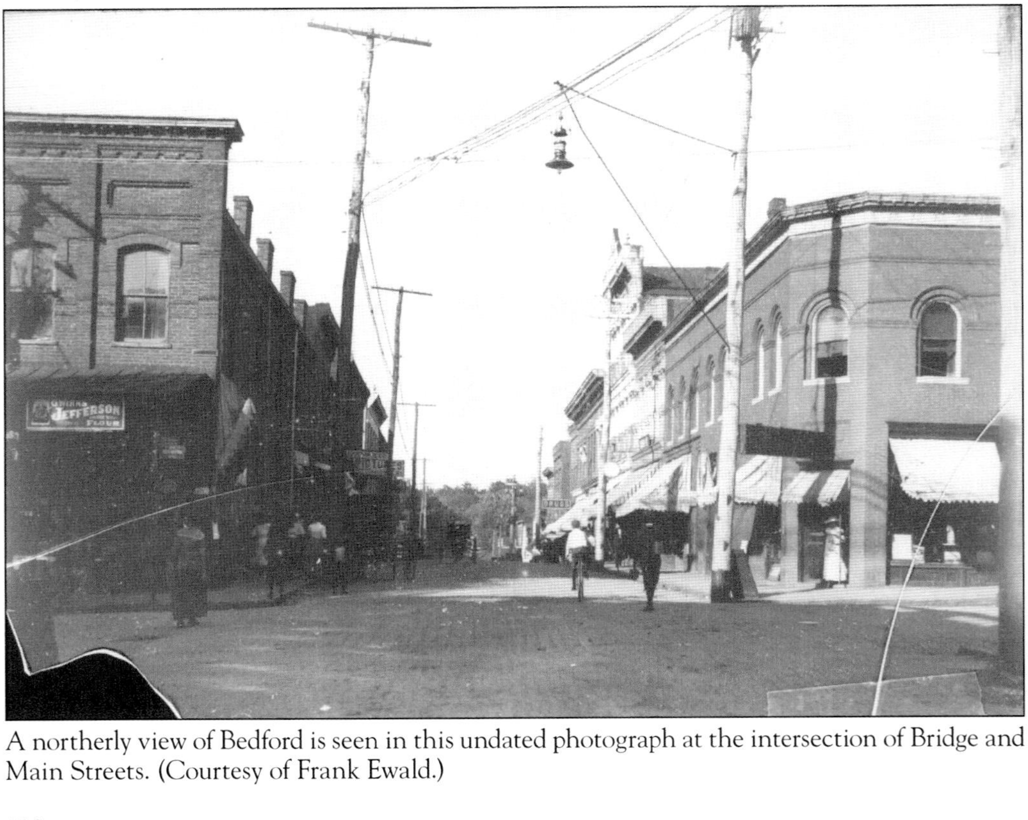

A northerly view of Bedford is seen in this undated photograph at the intersection of Bridge and Main Streets. (Courtesy of Frank Ewald.)

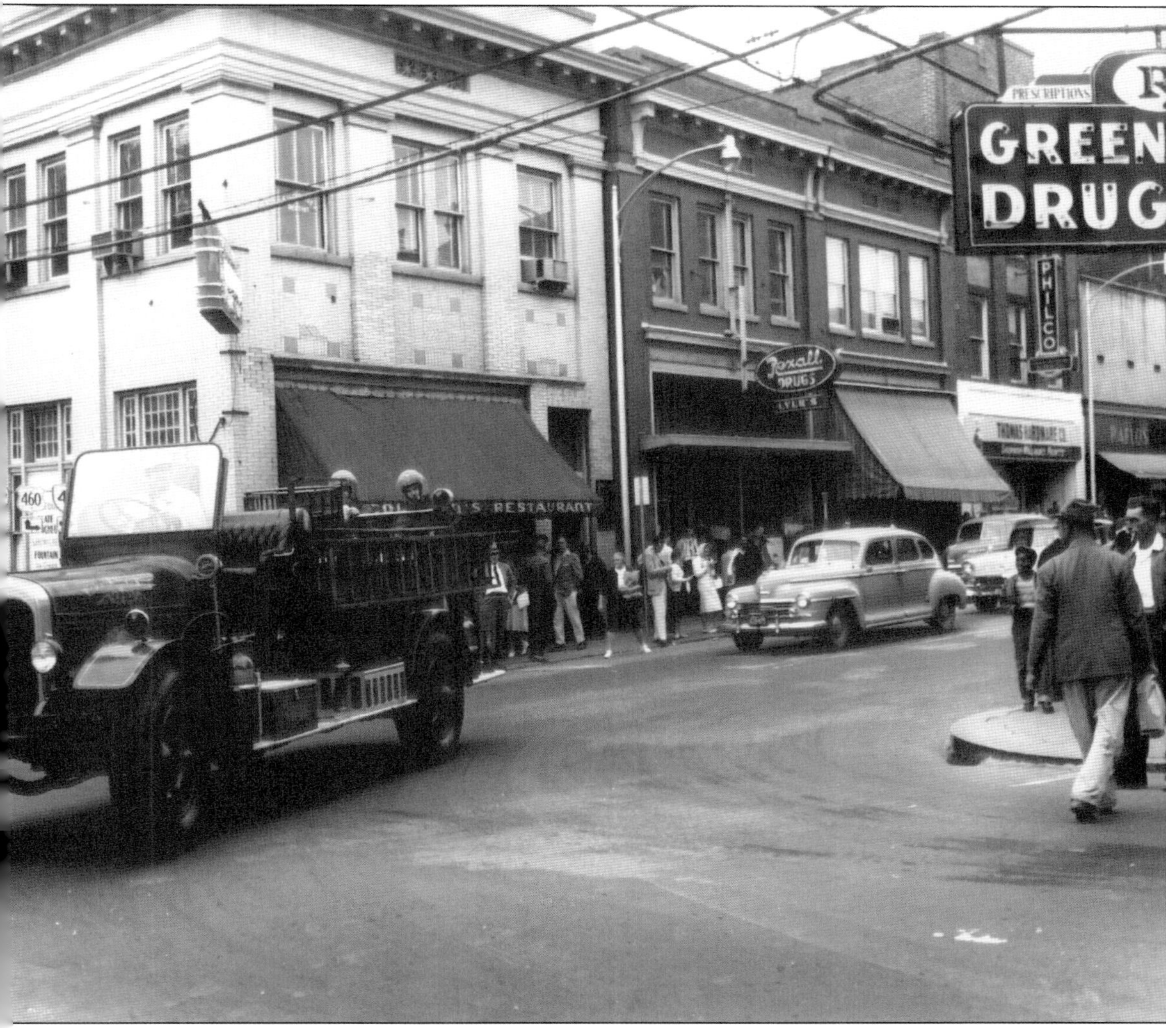

The same fire truck from an earlier photograph is now headed in the opposite direction. Its driver is now turning from South Bridge Street onto East Main Street. The Green's Drug Store on the right corner was a staple in the community for years and was often used by Dr. Sam L. Rucker Jr. of Moneta for his patients. (Courtesy of Frank Ewald.)

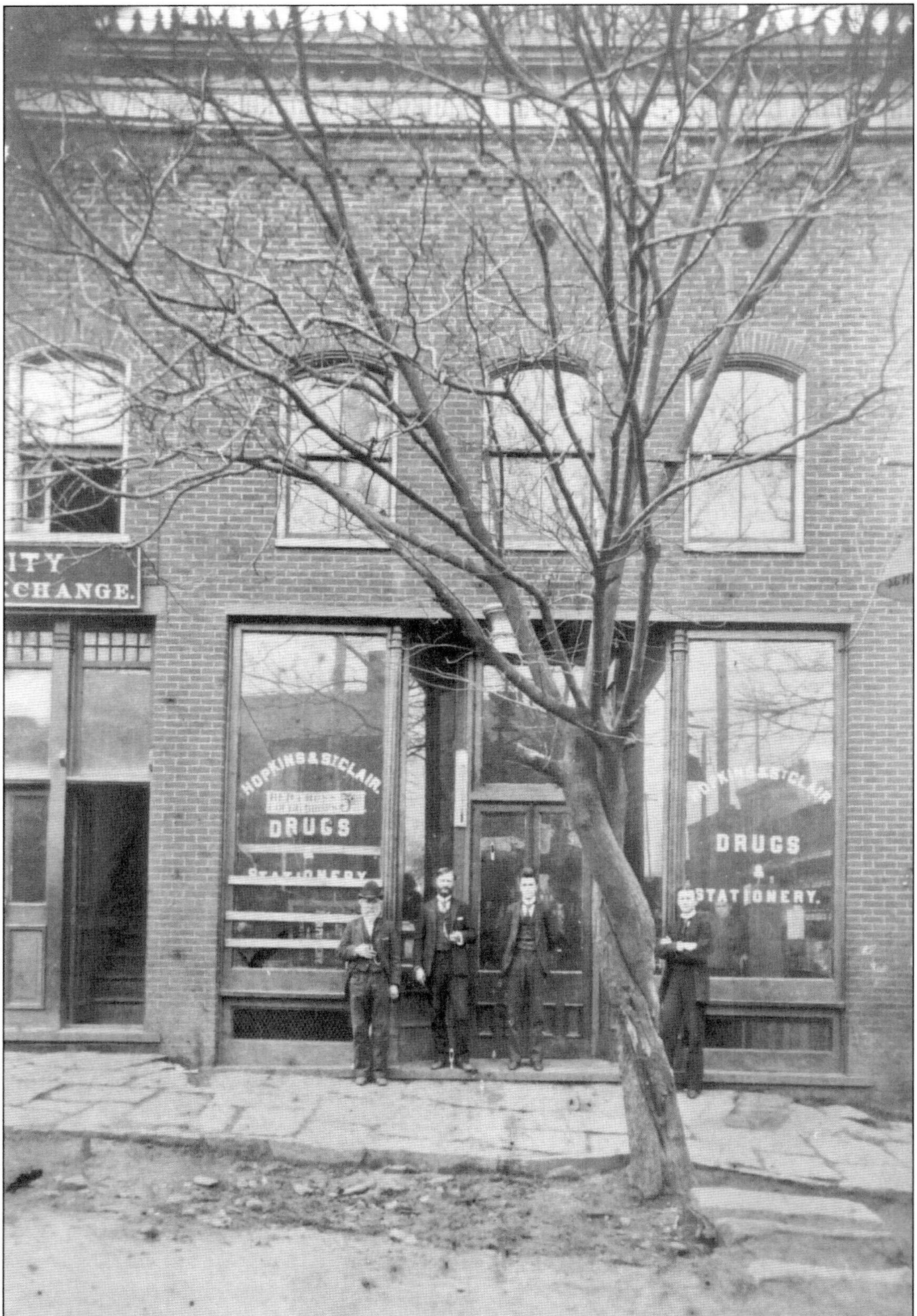

On the southeast corner of Main and Bridge Streets in Bedford, four men pose outside Hopkins and St. Clair Drugs and Stationery in 1890. (Courtesy of Frank Ewald.)

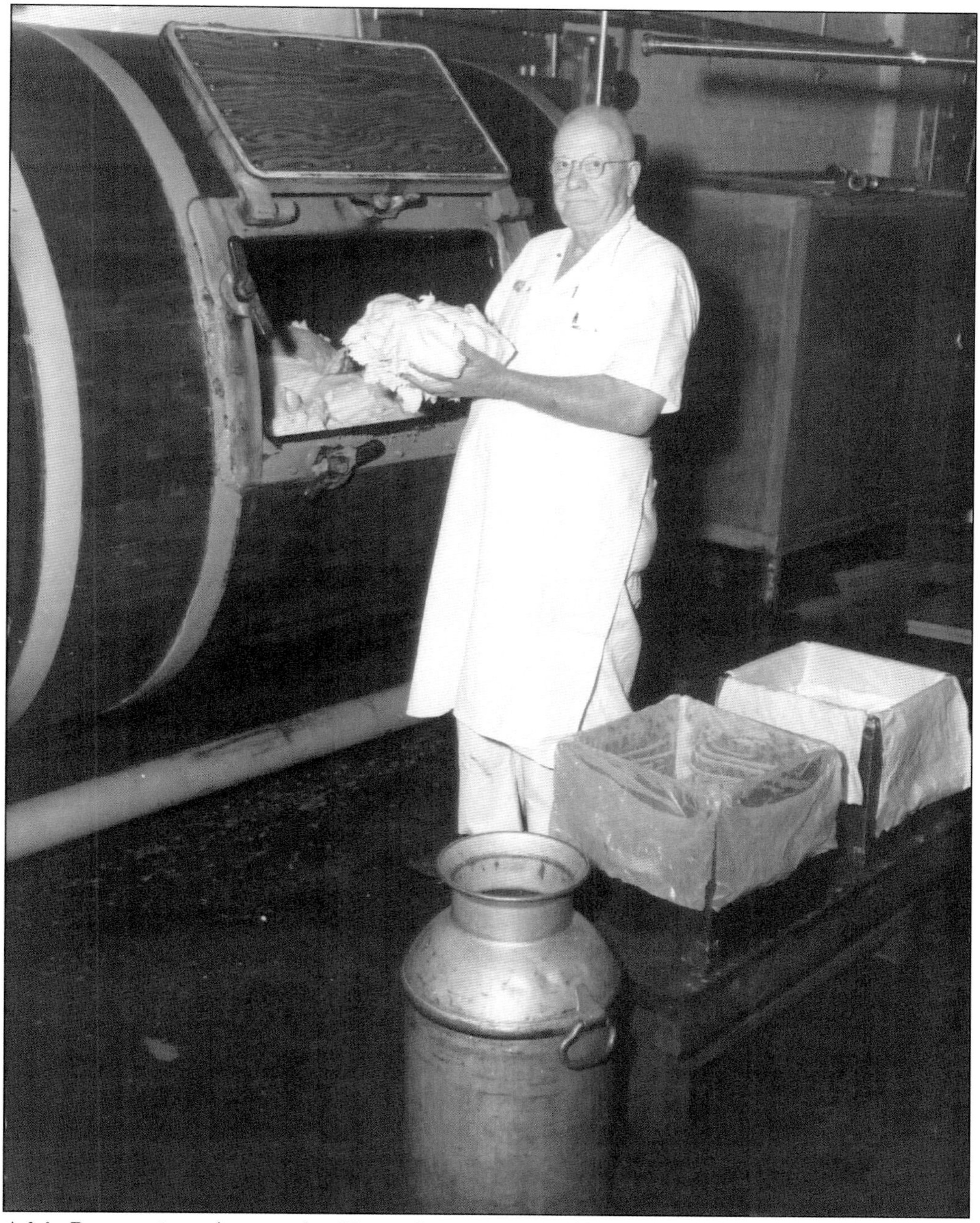

A Mr. Braason is working in the Clover Creamery in Bedford, probably during the late 1950s, in this photograph. (Courtesy of Frank Ewald.)

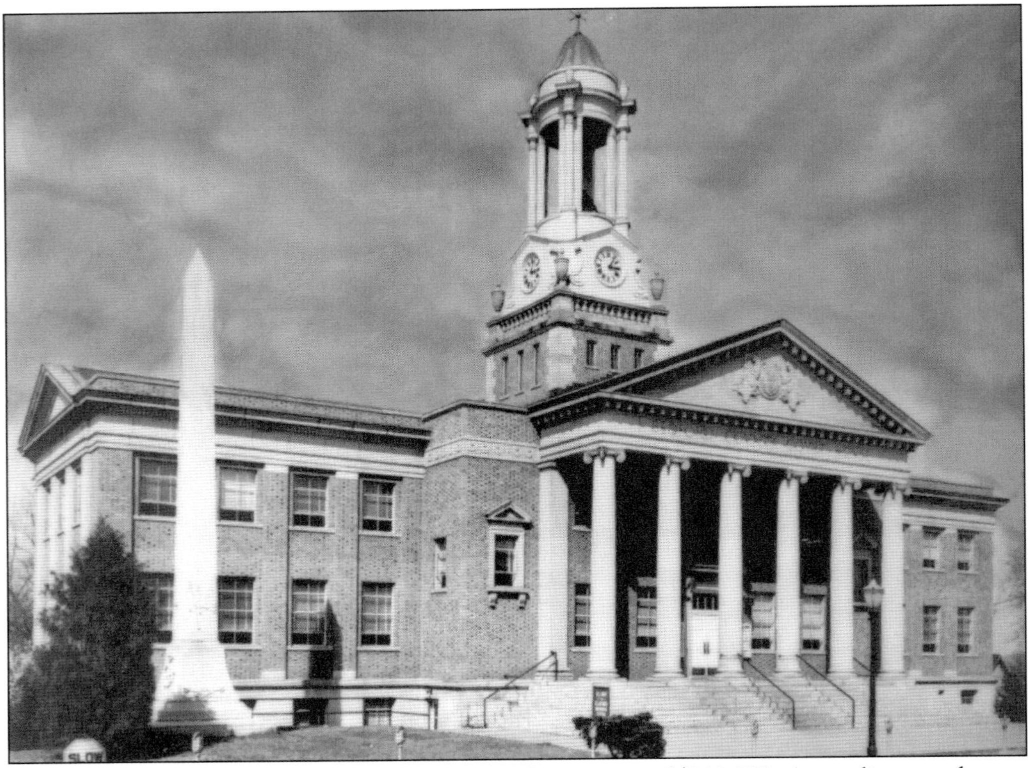

The Bedford County Courthouse was built in 1930 at a cost of $212,000. An earlier courthouse, which was erected in 1834, once stood on the same site. (Courtesy of Frank Ewald.)

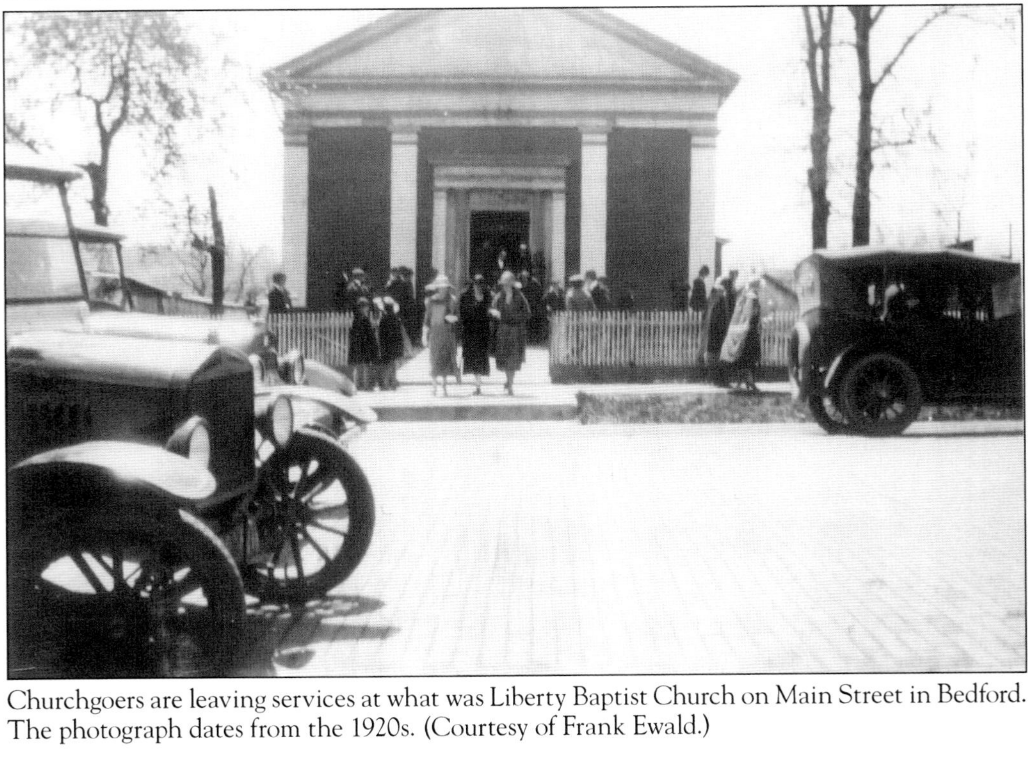

Churchgoers are leaving services at what was Liberty Baptist Church on Main Street in Bedford. The photograph dates from the 1920s. (Courtesy of Frank Ewald.)

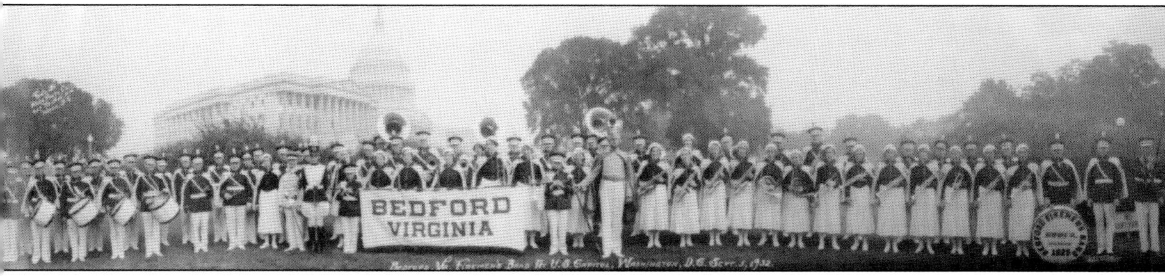

In this undated and unidentified photograph, a band from Bedford poses near the U.S. Capitol. (Courtesy of Frank Ewald.)

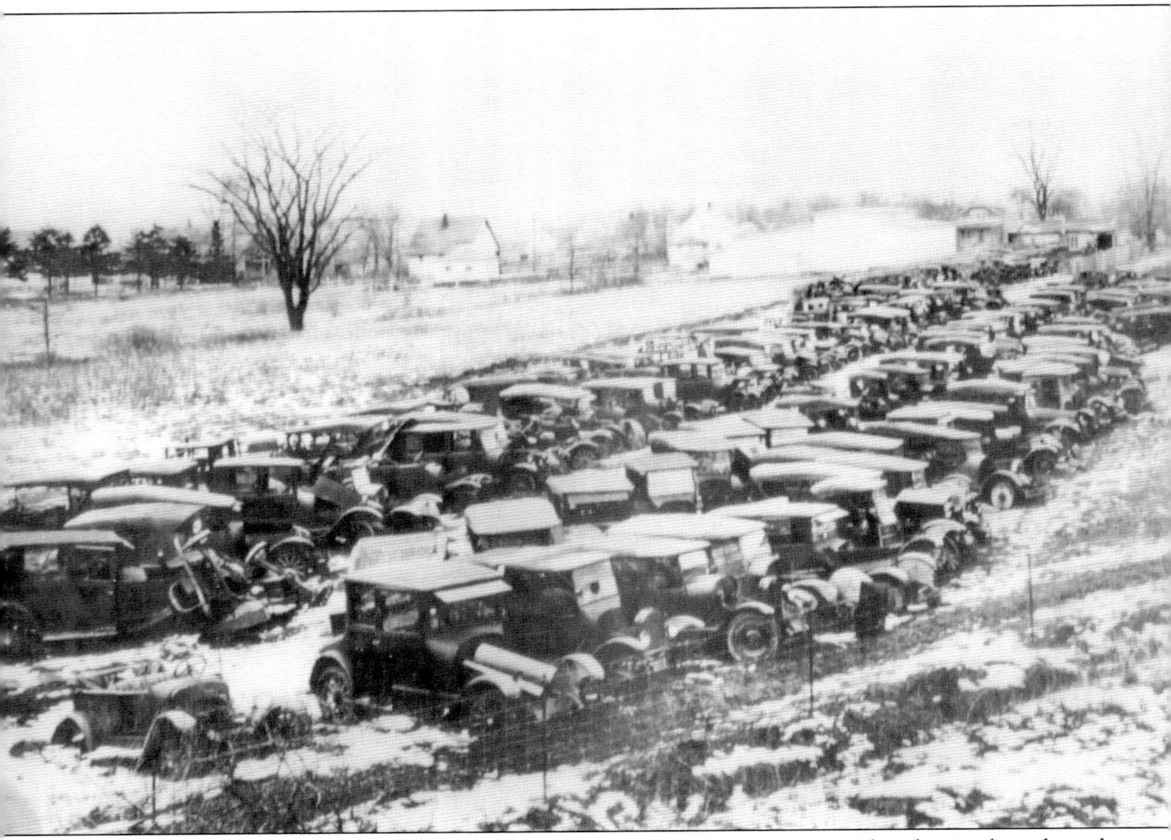

One person's junk is another person's treasure. Though it was considered a junkyard at the time, this field of jalopies in Bedford would be considered a gold mine now. (Courtesy of Frank Ewald.)

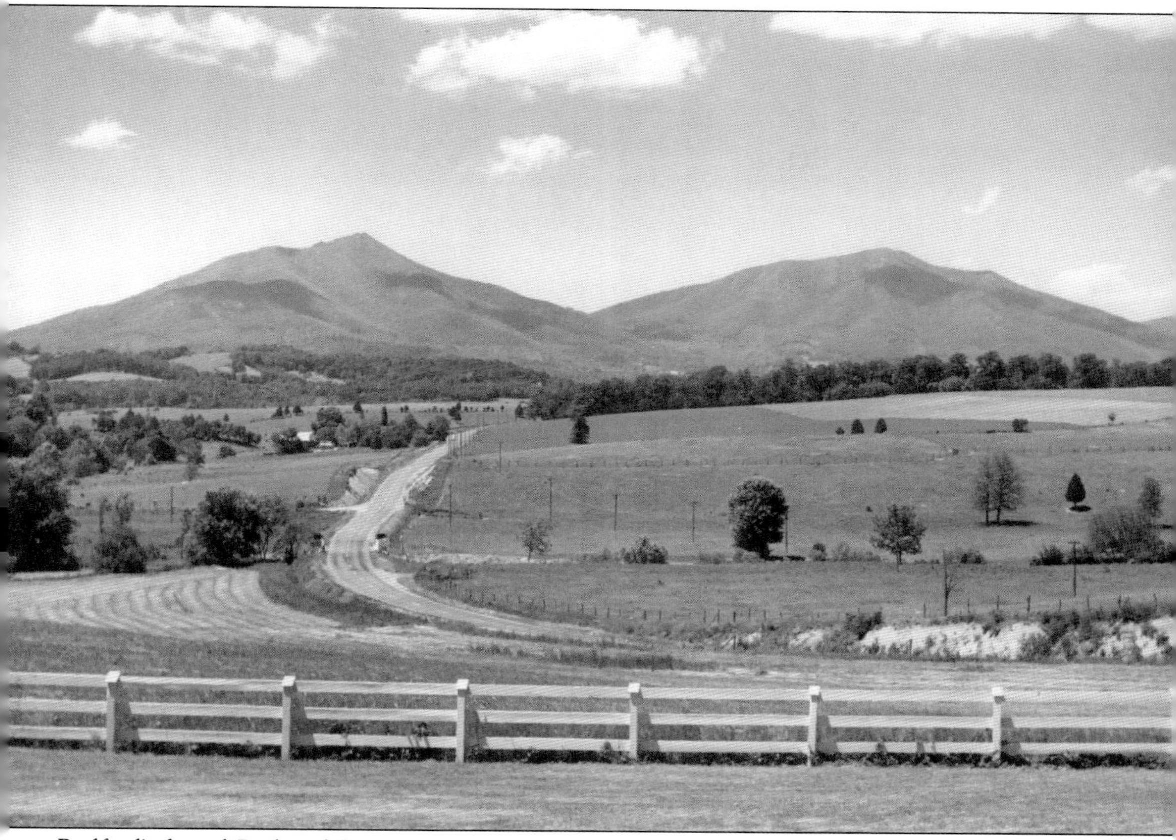

Bedford's famed Peaks of Otter are captured in this idyllic 1950s photograph. A June 6, 1929, *Bedford Bulletin* article says, "It is not known at what date the Peaks were 'discovered.' They had been known and visited for a time long antedating the Revolutionary War, visited by all those, who living near, possessed the hardihood to brave the wilds of the hills. They were a point of pilgrimage for the wealthy Virginia planters who came with their entourages of family and servants to climb the topmost rocks." (Courtesy of Frank Ewald.)

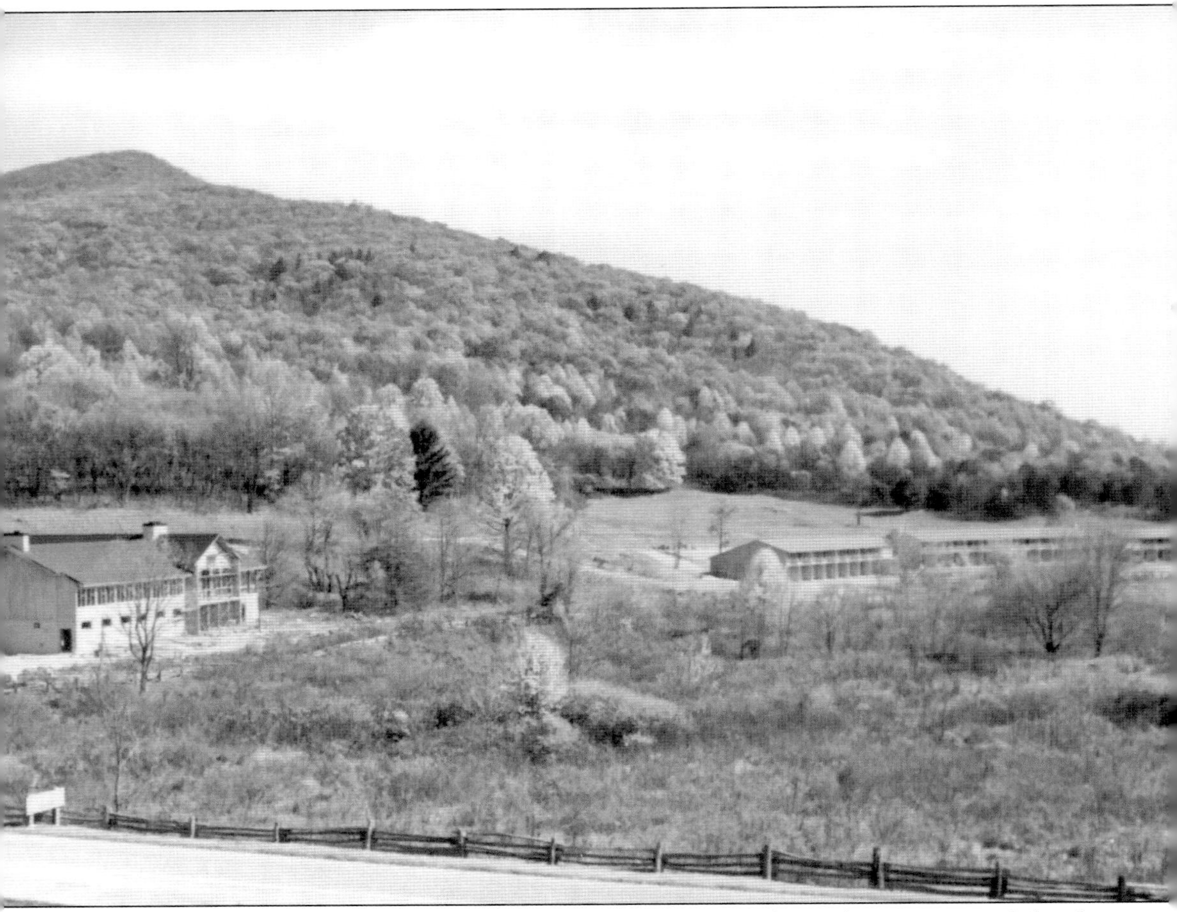

The Peaks of Otter Lodge is located on the Blue Ridge Parkway, between two of the three mountains that comprise the lovely Peaks of Otter. This inviting photograph shows the facility shortly after its completion in 1964. The Web site invites visitors to take in the lovely view and the wildlife, including an occasional bear. (Courtesy of Frank Ewald.)

Nine

Hollywood Comes to Town

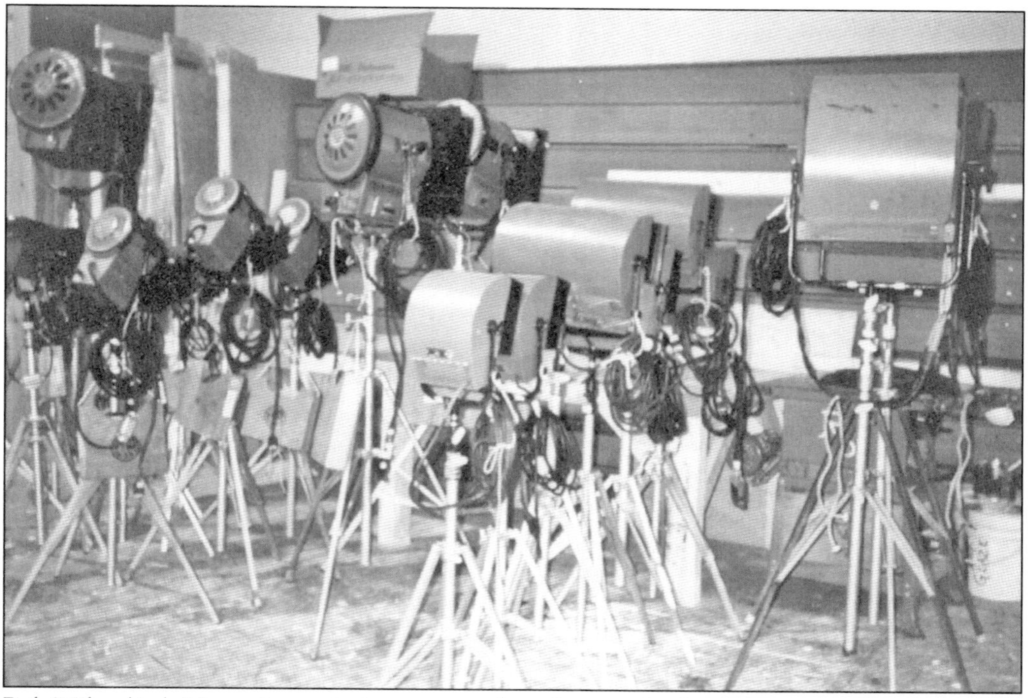

Bob Wiley (Bob Murray), a reclusive patient of psychiatrist Dr. Leo Marvin (Richard Dreyfuss), learns that he has been referred to another doctor. Frantic, he follows the Marvins on their vacation trip to Lake Winnipesaukee, New Hampshire, and all kinds of comedic trouble ensued. Released in 1991, *What About Bob?* generated much excitement among Bedford County residents. It also was evidence that life here had changed—dramatically. In this photograph, lights are being stored in the Bedford Armory for future on-premise film shooting. (Courtesy of Mike and Leslie Mehaffey.)

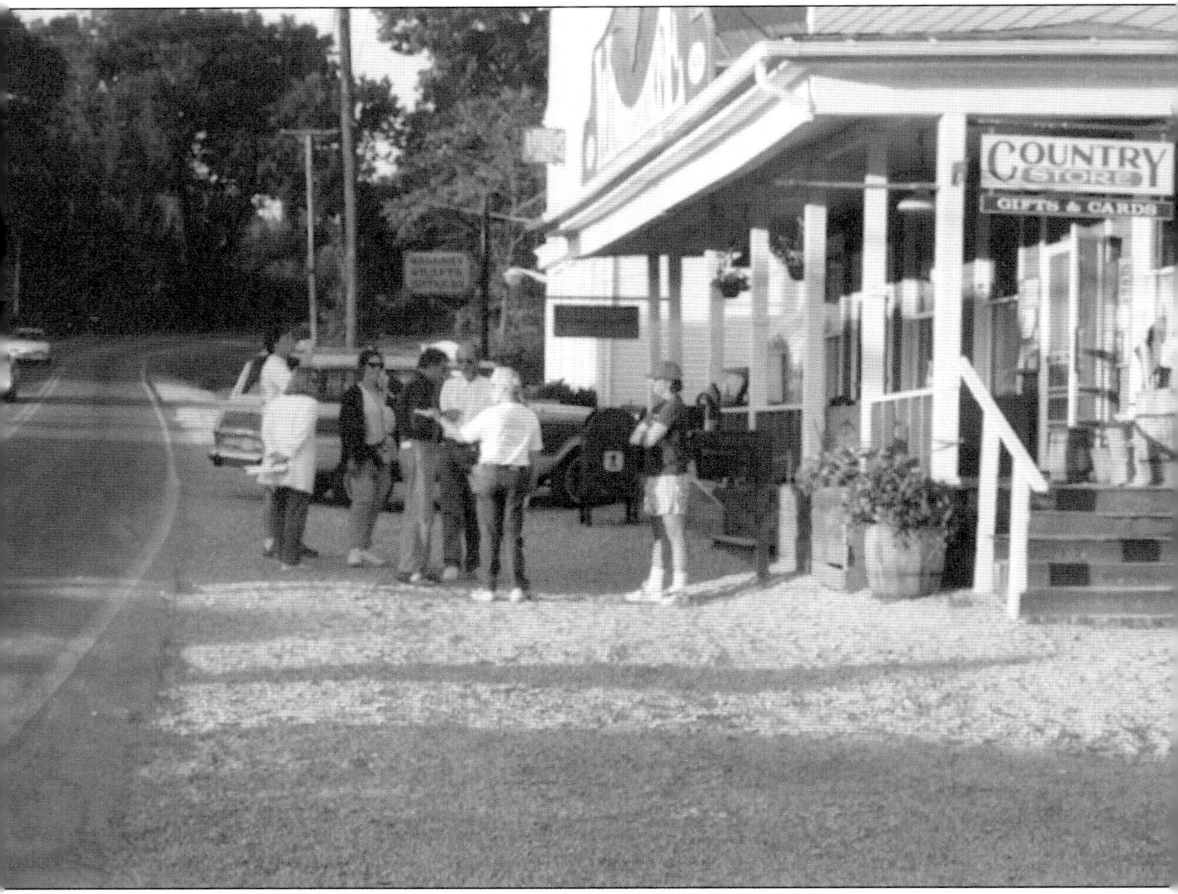

Richard Dreyfuss (center, in white T-shirt and blue jeans) and Bill Murray (right) discuss a scene during filming of *What About Bob?* in front of the William Meador store in Moneta. (Courtesy of Mike Mehaffey.)

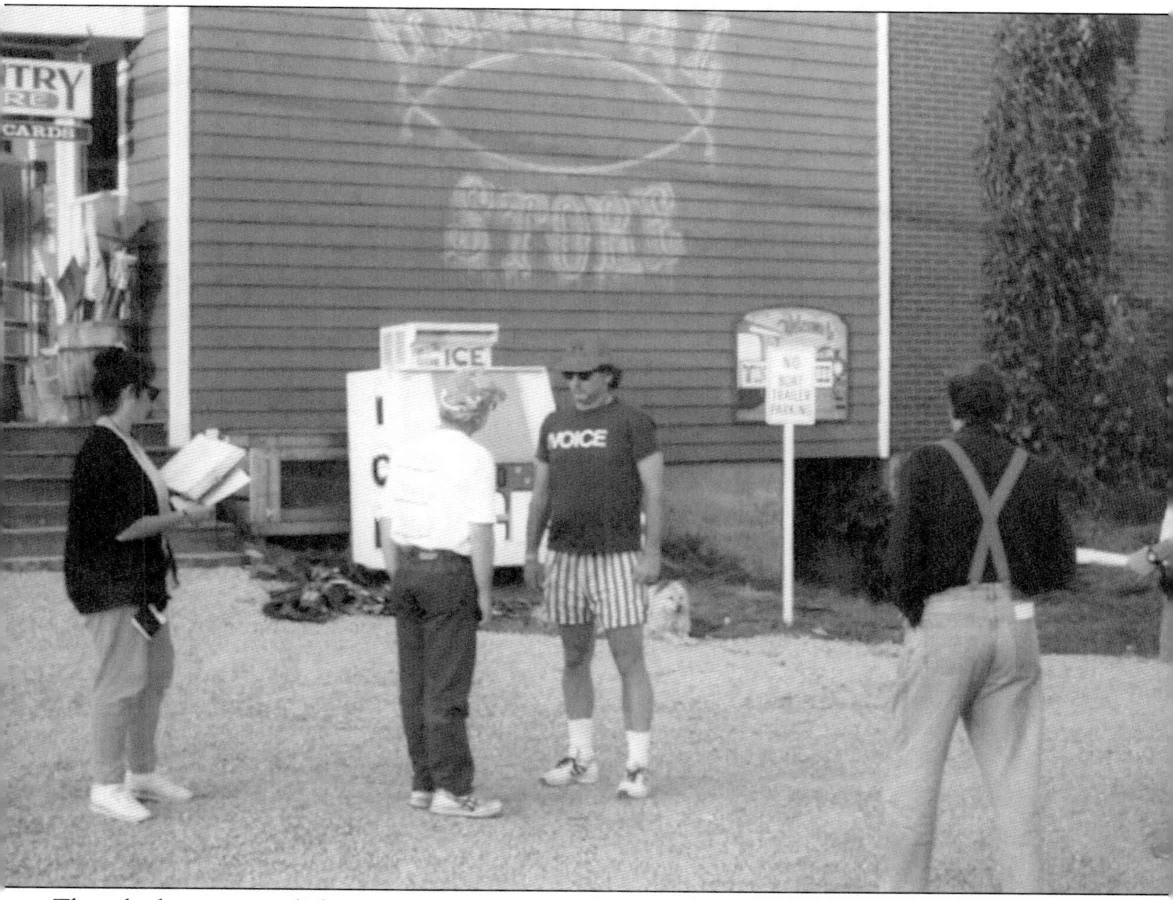

Though there are no lights or cameras present, there is plenty of action in this photograph. Dreyfuss and Murray have finished their scene discussions and are now in rehearsal outside the general store in Moneta. (Courtesy of Mike Mehaffey.)

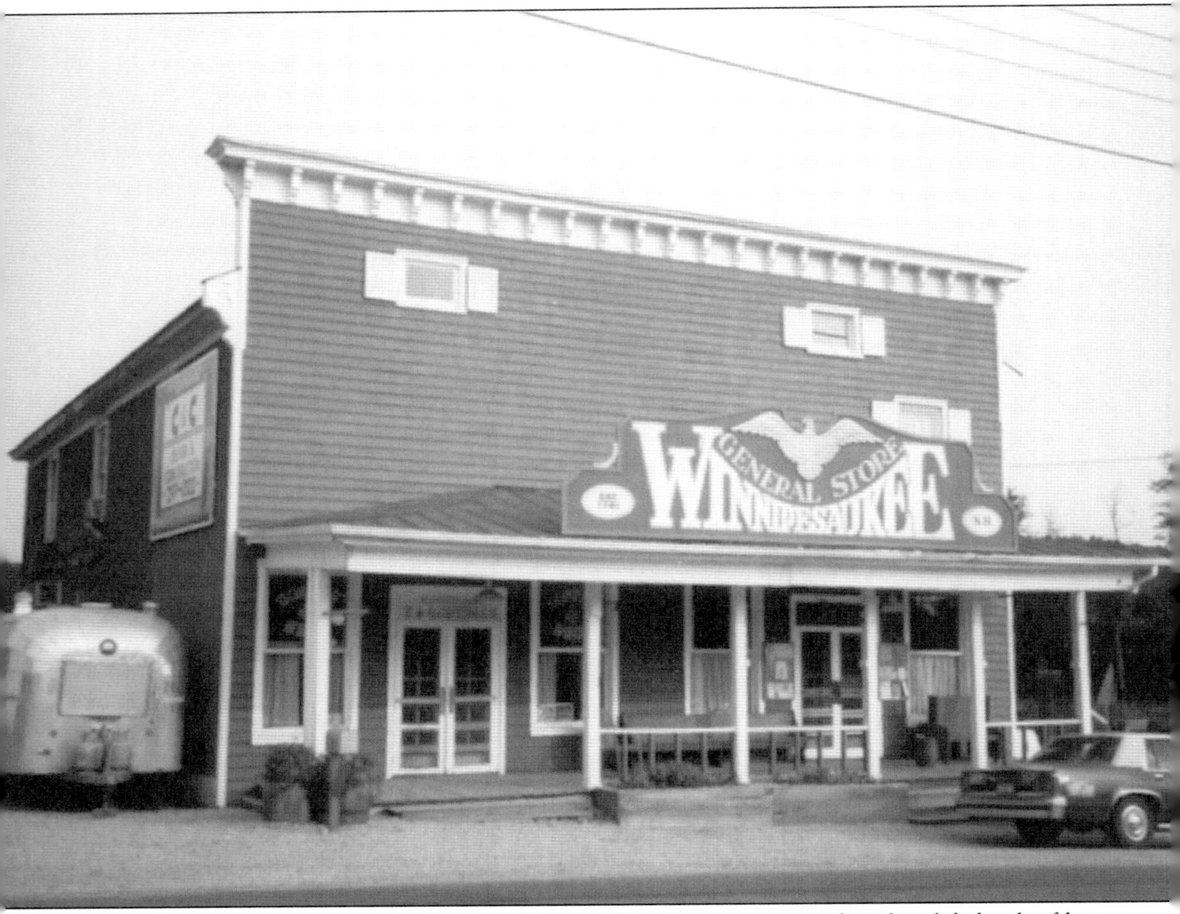

Touchstone Pictures gave the William Howard Meador store a complete face-lift for the filming of the movie. The notable improvements included a refurbished front porch with steps that no longer sagged, wood painted a relaxing shade of blue, and attractive flowers planted in barrels by the front steps. The building served as the Lake Winnipesaukee General Store and the coffee shop. (Courtesy of Mike Mehaffey.)

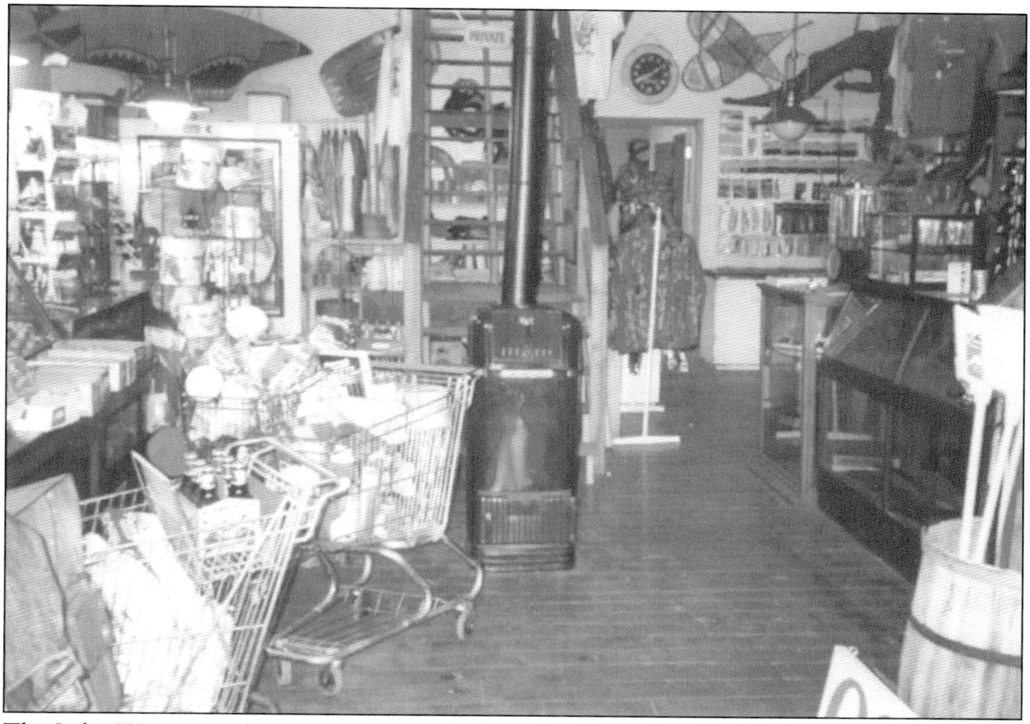

The Lake Winnipesaukee General Store had a cozy, authentic charm, right down to its Warm Morning stove. (Courtesy of Mike Mehaffey.)

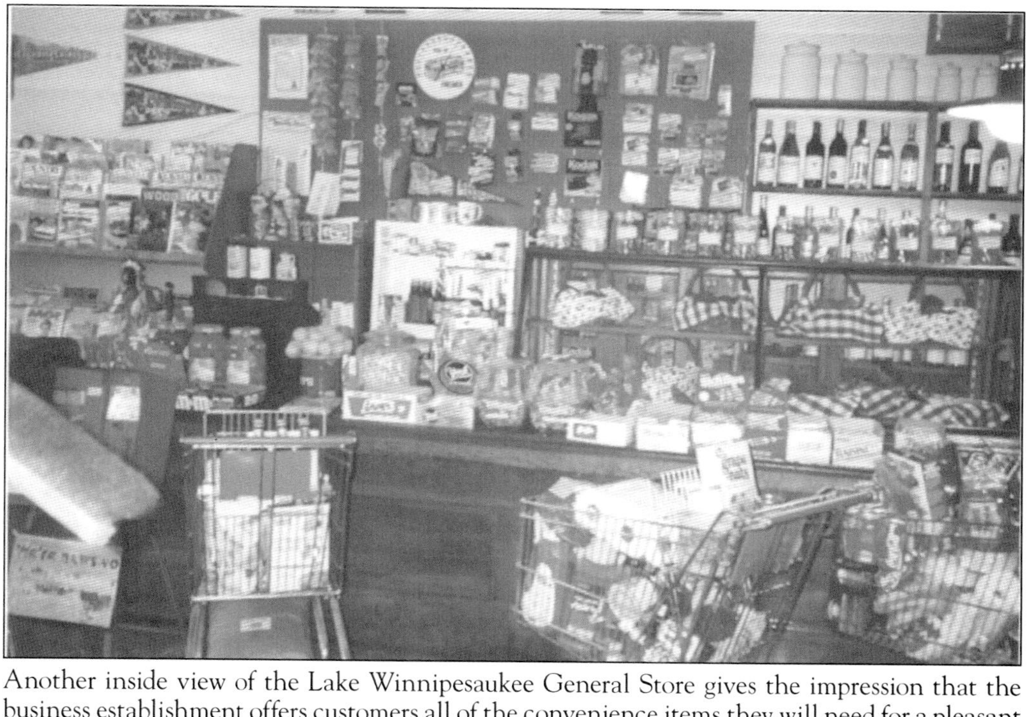

Another inside view of the Lake Winnipesaukee General Store gives the impression that the business establishment offers customers all of the convenience items they will need for a pleasant vacation. The ambience of a bona fide, early-20th-century general store adds to the experience for the shopper and the moviegoer, too. (Courtesy of Mike Mehaffey.)

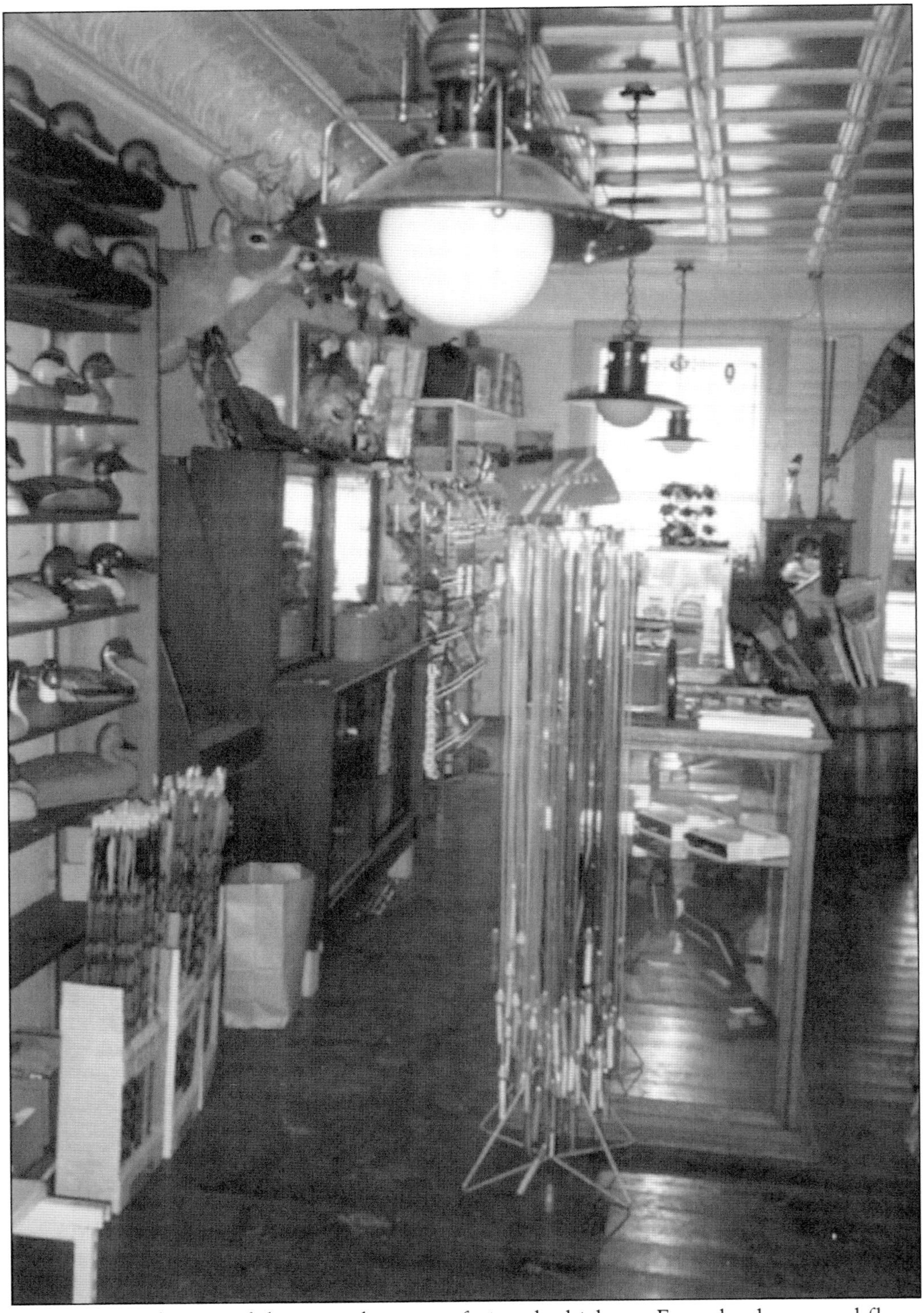

This is an inside view of the general store set facing the highway. Even the dusty wood floors were given a proper shine for the movie's scenes. (Courtesy of Mike Mehaffey.)

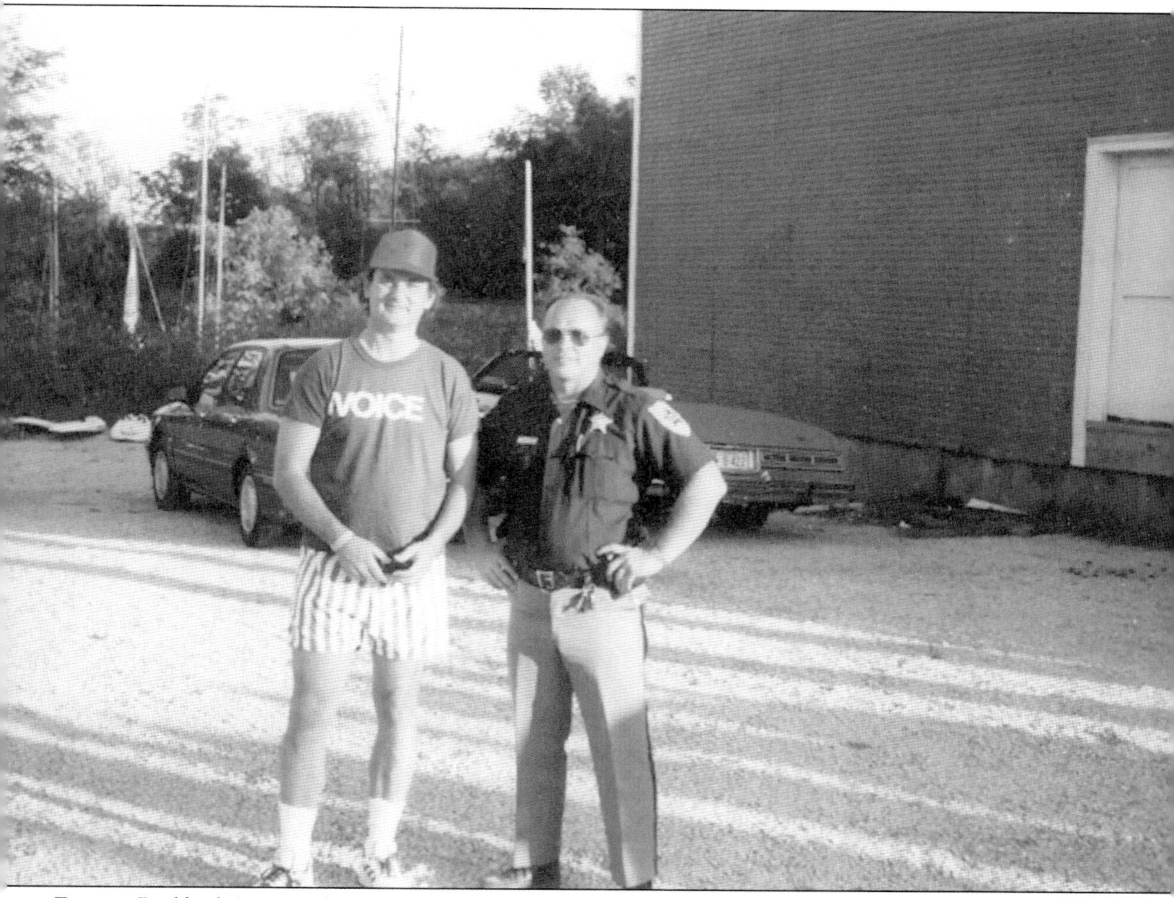

Former Bedford County deputy sheriff Mike Mehaffey poses for a photograph with Bill Murray during a break in the shooting of *What About Bob?* (Courtesy of Mike Mehaffey.)

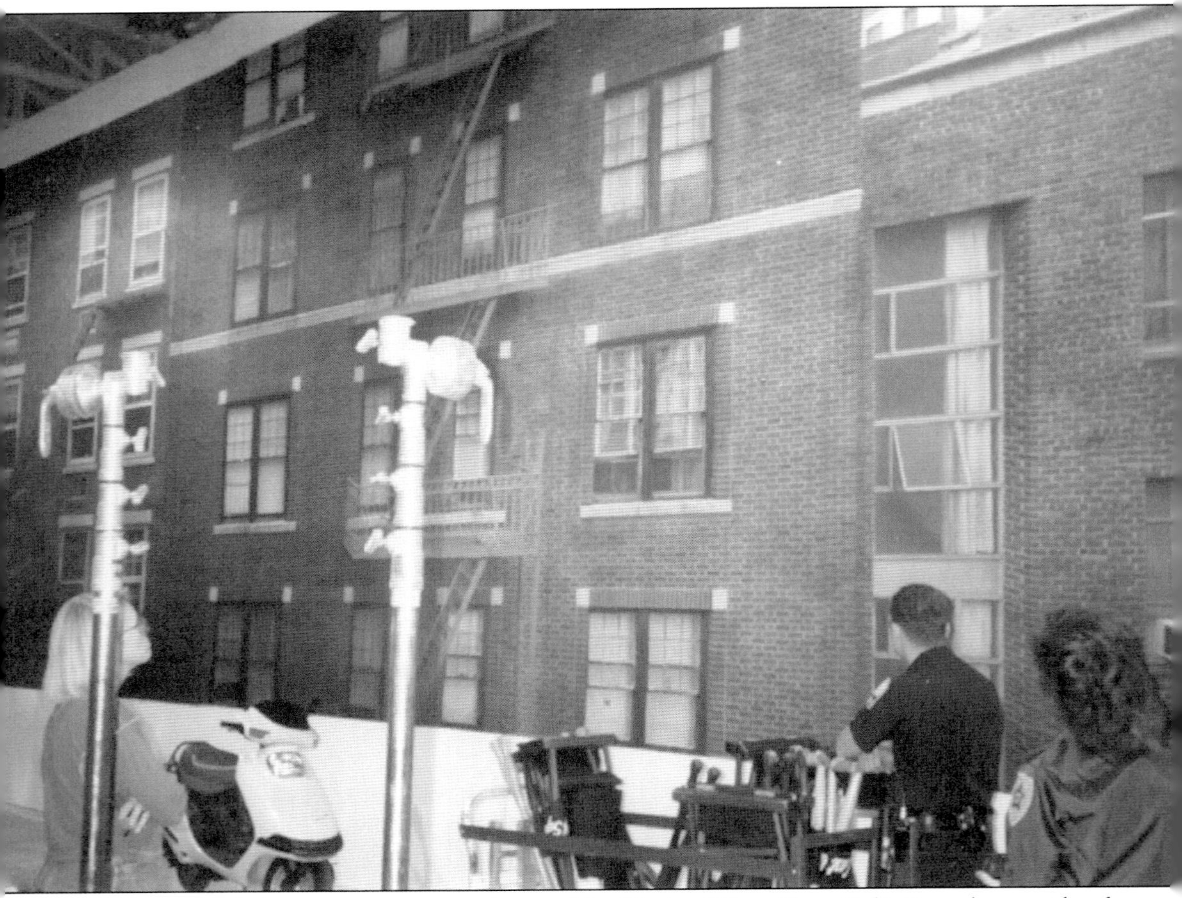

The Bedford Armory interior was also used for filming *What About Bob?* Here the outside of a city building has been constructed and is ready to be used in a scene. Note the size proportions of the set scene with the bystander on the left, Leslie Mehaffey. (Courtesy of Mike Mehaffey.)

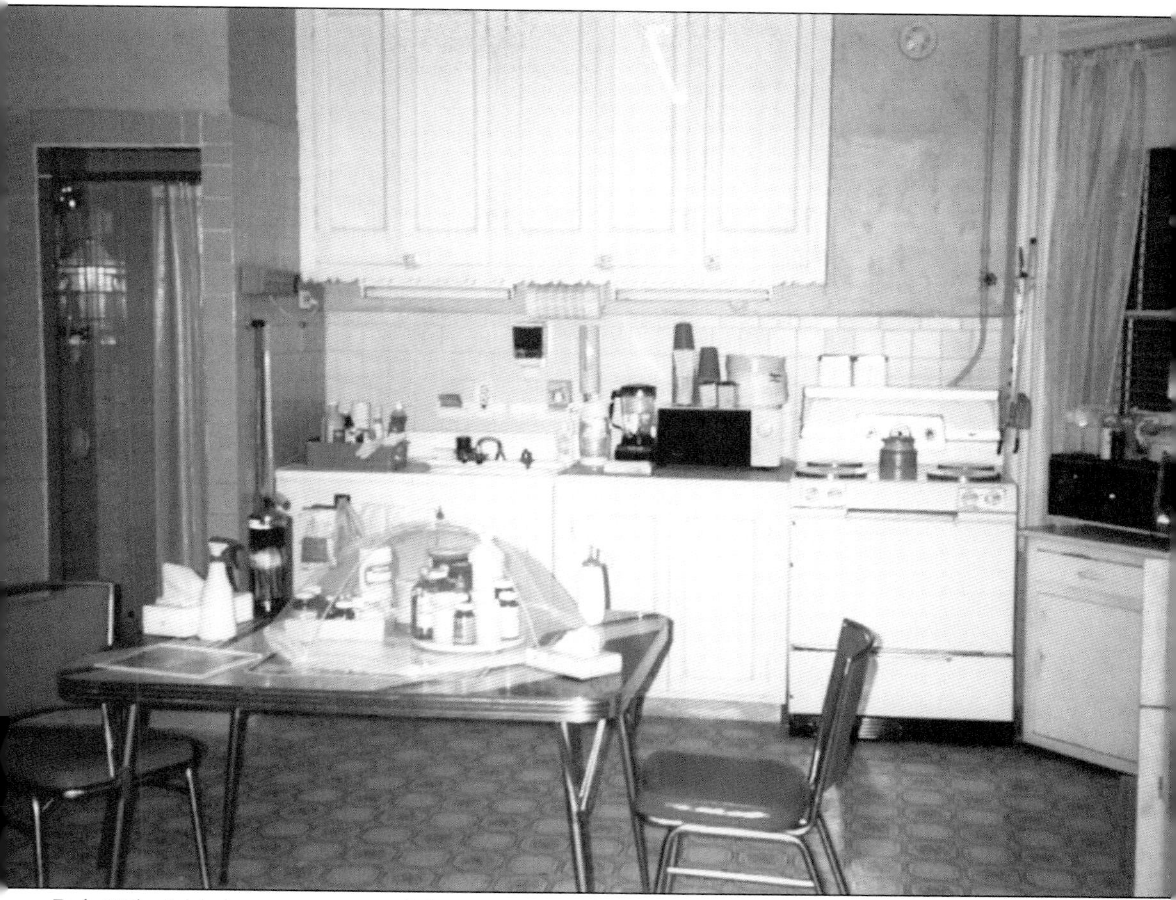

Bob Wiley's kitchen is empty, and this may mean that he is now in pursuit of the Marvin family at Lake Winnipesaukee. Like the building exterior in the preceding photograph, the interior of Bob's apartment was built inside the Bedford Armory. (Courtesy of Mike Mehaffey.)

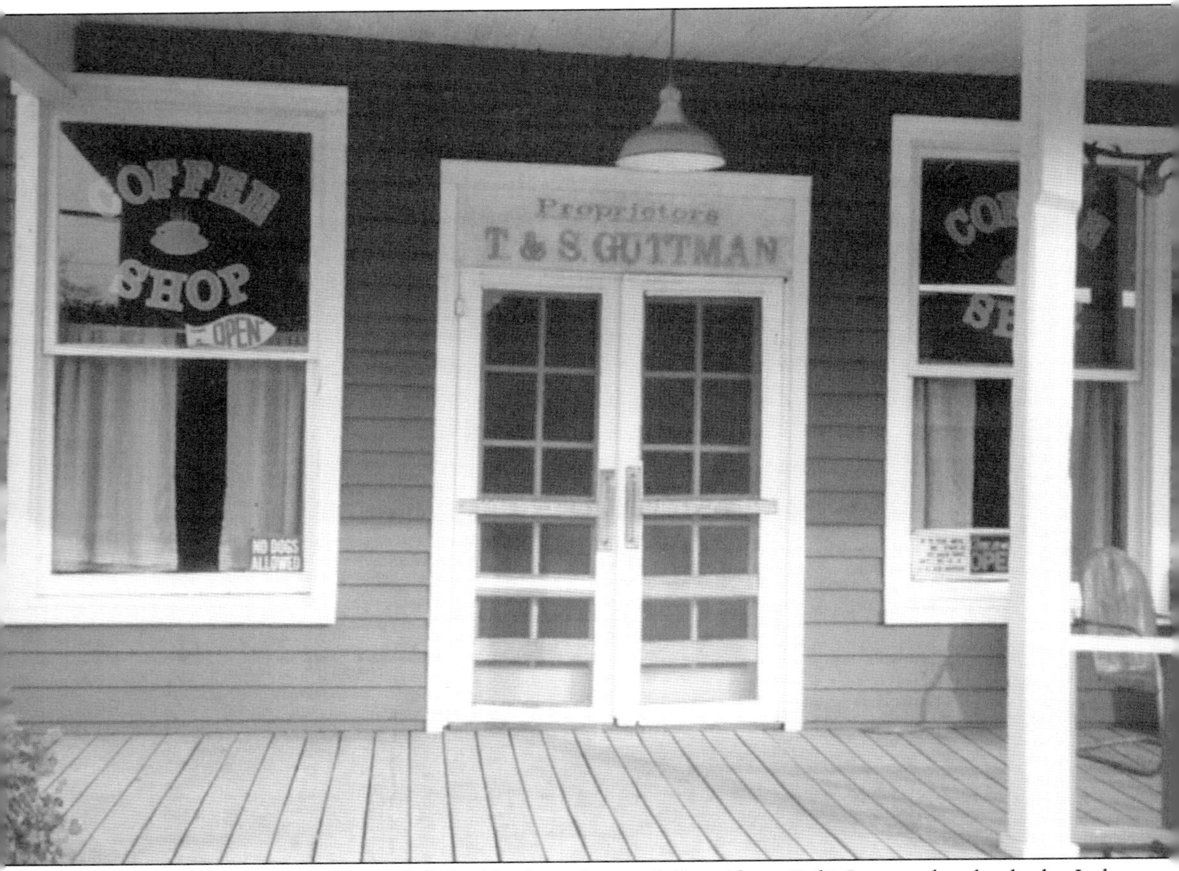

The Howard Meador store did double-duty during *What About Bob?* It served as both the Lake Winnipesaukee General Store and as a coffee shop. (Courtesy of Mike Mehaffey.)

Harold's cow visited the yacht club

Dear Mrs. Jackson-Clause,

Folks who say nothing's happened down our way since the Huddleston bridge went south weren't here June 22. I'll tell you, we had a time. It'll go down in White House community history as the day Harold's cow went to the yacht club. Yes ma'am, that's no misprint. Tuesday, a week or so back, Harold Martin's cow visited Smith Mt. Yacht Club. Of course, there's more to it than that. First of all, there were two cows involved, but only one made it to the docks. The other had her own adventure in the road out of sight. It wasn't but a minute or so before Ferris Blankenship called saying they'd struck a lope through his garden.

Now Harold, who'd turned these beauties out that morning and watched them disappear like two wild jackalope, was already on the road in pursuit. He told me later, "Glenn, your boy said it perfect when I asked him about those cows. He said, 'Yes, Mr. Martin, I seen them and they were escaping!' That's exactly what they was doing — escapin'."

Well, Mrs. J-C, to make a long story short, those critters parted ways somewhere beyond Ferris, because the one went to Clyde Saunders', double-backed on the Stony Mt. road, and passed the gum tree making for the railroad. Whether she expected to take train there or just wanted a nice ramble in the country, Lord knows, but she came to rest—according to Harold—in Calvin Willard's pasture.

The other beast, was a heap of trouble. You can imagine Harold's feelings when the yacht club called to report his cow nosing about the fleet. So, he and Ann struck out down there, wondering whether that four-footed heathen was trying to book passage back to Angus shire, but worried more what an animal valued at a few hundred dollars might do clambering on board a schooner worth five figures. Anyway, when they got there, the

what some of our neighbors have already labeled the"Great Escape." Here's how it happened.

My boy, James Frye was standing barefoot on his porch enjoying a cup of tea by the dawn's early light, when suddenly these cows burst out of the thicket below our tractor shed like two stegosauri from a scene in JURASSIC PARK. Since James's house is the farthest point on our farm from Harold, he naturally thought they were ours, but time he'd laced his boots, they'd lit out

cow had hit for the Shoprite corner where she finally settled down with Aubrey Bobbitt's herd, plumb wore out, after a tour of about thirty miles in what seemed like thirty minutes.

Old Harold was powerful relieved I'll tell you. He didn't even seem too anxious to get the cows back, and I can't say I blame him. It's tough on a semi-retired cowboy to chase all over the country following what VPI calls "bovine perenigration," particularly when the bovine perenigrate to such places that a body can't decide whether to go on the truck or the car and whether to wear coveralls or a blazer and white ducks.

I'm right worried myself, about the tastes of that one cow. Seems she's too much into the "lake life" to make a good herd animal. No telling where she'll want to go next. Like as not she might stroll over the bridge to Turtle Island or walk in King's Point Pub, order a pint of Killian's, and ask Dave Owens to sing, "Mamas, Don't Let Your Babies Grow Up To Be Cowboys." I'll tell you, Mrs. J-C, I'm giving all my friends who play golf over to Mariner's Landing this warning: if you see a cow coming up the fairway behind you, don't panic; just mark your ball, let her play through — and call Harold.

That's it from here, Mrs. J-C, thanks for your support.

Glenn M. Ayers
Route 1, Box 56
Moneta, Va. 24121

This letter to the editor in a mid-1990s edition of the *Bedford Bulletin* describes a comedic yet happy marriage of a resort community with an old-time farming community. Like its real-life counterpart,though, this marriage has had its ups and downs, too. Hopefully, there will be more hilarious high jinks in early-21st-century Bedford County. If so, the area might produce an even funnier movie than *What About Bob?* (Courtesy of the author.)